WITHDRAWN
UTSA LIBRARIES

Polymers for Engineering Applications

Raymond B. Seymour
Distinguished Professor
Department of Polymer Science
University of Southern Mississippi

 ASM INTERNATIONAL

Library of Congress Catalog Card No.: 87-70003

ISBN: 0-87170-247-9

SAN: 204-7586

Editorial and production coordination by
Carnes Publication Services, Inc.

PRINTED IN THE UNITED STATES OF AMERICA

Preface

In 1947, John Gunther wrote, "This is a steel age." Yet the sale of steel by the major American manufacturers has decreased by 40% since Gunther bragged about rolling over 90 million tons of steel ingots a year.

In a 1986 report titled "Up From the Ashes," Donald Barnett and Robert Crandall stated that the "Age of Steel" was fading fast. As an illustration, in the 1960s, employment in the steel and hotel industries was about equal, but the hotel industry currently employs five times as many workers as the steel industry, and this gap appears to be widening.

Of course some, like John Stroymeyer in his book *Crisis in Bethlehem,* place the blame for the decline of the American steel industry on corruption of management and labor. But it is important to recognize that we are now living in a new age of *materials,* in which engineers and designers are in a position to select the material that is best suited to a specific application, whether the material is a steel or a plastic or a ceramic.

While steel is far from becoming dispensable, it does have to compete with materials that were virtually unknown to Gunther in the 1940s. Today the engineer's or designer's choice must be based on a thorough knowledge of the science and technology of both classic and new materials, and polymers undeniably account for a large percentage of these new materials.

This book has been written in an attempt to present specific information about polymers for engineering applications. My hope is that this information will be helpful to both engineers and designers in their search for the most appropriate material for each application.

RAYMOND B. SEYMOUR
Hattiesburg, MS

Contents

5   High-Performance Fibers   65

9 Moderately High-Performance Polymers   115

10 Engineering Polymers   127

11 Physical and Chemical Testing of Polymers   151

12 Terms and Symbols, Trade Names, and Bibliography   159

Appendix: U.S./SI Units: Definitions and Conversions 177

Index   181

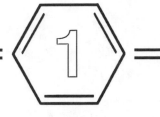

Important Polymer Concepts

Polymers are relatively new materials that differ from traditional materials of construction, such as metals and ceramics, in their molecular structures. However, many of the concepts applied to these traditional materials of construction may also be applied to polymers; and, of course, some of the traditional materials, such as wood and cotton textiles, are polymeric materials.

1.1 Polymers vs. Metals

All polymers, whether organic or inorganic, consist of extremely large molecules, called macromolecules, in which the atoms in the polymer chain are held together by covalent bonds, as shown in the segment of polyethylene below. In contrast, as depicted in Fig. 1–1, metal cations (+) are held together by a cloud of loosely held valence electrons (−).

$$-\ :\overset{\displaystyle \overset{H}{:}}{\underset{\displaystyle \underset{H}{:}}{C}}:\overset{\displaystyle \overset{H}{:}}{\underset{\displaystyle \underset{H}{:}}{C}}:\overset{\displaystyle \overset{H}{:}}{\underset{\displaystyle \underset{H}{:}}{C}}:\overset{\displaystyle \overset{H}{:}}{\underset{\displaystyle \underset{H}{:}}{C}}:\ -$$

Many of the so-called metallic properties, such as luster, electrical conductivity, malleability, and ductility, are a function of the presence of fluid electrons distributed about the cluster of spherical cations in metals. Likewise, many of the so-called nonmetallic properties of organic compounds (including

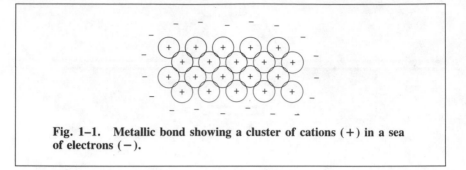

Fig. 1–1.   Metallic bond showing a cluster of cations (+) in a sea of electrons (−).

macromolecules), such as low specific gravity, low conductivity, transparency, and resistance to corrosion, are a function of the covalently bonded electrons.

In addition to possessing the characteristic properties of small organic molecules, such as octane (C_8H_{18}), polymers also possess properties characteristic of their large size; that is, long chains that can be entangled with and attracted by other long chains. Many of the differences among plastics, fibers, elastomers, coatings, biopolymers, and other materials depend on the functional groups present on the polymer chains and the extent of interaction among these chains.

1.2 Covalent Bonding in Polymers

In contrast to a simple organic hydrocarbon compound such as octane, which consists of eight catenated carbon atoms covalently bonded in a short continuous chain, a hydrocarbon polymer, such as linear polyethylene (LDPE), will consist of 100 or more catenated carbon atoms covalently bonded in a long continuous chain, as shown by the following formulas in which the covalent bonds (:) will be shown as single bonds (—):

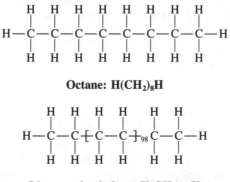

Octane: $H(CH_2)_8H$

Linear polyethylene: $H(CH_2)_{200}H$

For purposes of simplicity, the true bond angles are not shown above. Actually, the characteristic bond angle for all carbon-carbon bonds in organic compounds is 109.5°. The length of the carbon-carbon bond is 1.54×10^{-8} cm (1.54 Å, 1.54×10^{-7} mm), but since the so-called linear chains have a zig-zag structure because of the characteristic bond angles, the distance between carbon atoms along the direction of the chain is 1.26×10^{-8} cm (1.26 Å, 1.26×10^{-7} mm). Thus, the full contour or stretched length of the linear polyethylene molecule shown above would be 1.52×10^{-5} mm.

1.3 Linear and Branched Polymers .

The polymer chains are able to twist and coil and are seldom stretched to their full contour length, because of the so-called free rotation about single carbon-carbon bonds. Since the size of the covalently bonded hydrogen atoms is insignificant compared to the total length of the covalently bonded carbon atoms in polyethylene, we may designate these macromolecules by a simulated continuous-line structure, as shown in Fig. 1–2.

In addition to the linear configuration, as shown in Fig. 1–2, polymer chains may have a branched structure, in which, as shown in Fig. 1–3, repeating units of the polymer molecule are covalently attached to the polymer chain or backbone, like the hydrogen atoms in linear polyethylene.

1.4 Polymers With Pendant Groups

Commercial high-density polyethylene (HDPE) is a linear polymer, whereas commercial low-density polyethylene (LDPE) is a branched polymer. Many polymers, such as polyvinyl alcohol (PVA), polyacrylic acid (PAA), polyvinyl

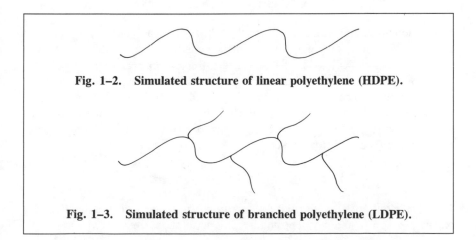

Fig. 1–2.   Simulated structure of linear polyethylene (HDPE).

Fig. 1–3.   Simulated structure of branched polyethylene (LDPE).

chloride (PVC), polystyrene (PS), polypropylene (PP), and polyacrylonitrile (PAN), have the functional groups OH, COOH, Cl, C_6H_5, CH_3, and CN in place of some of the hydrogen atoms on the chain. However, these are present in the repeating units made up of the monomer units in the polymer chain and are called pendant groups, not branches. Typical repeating units with pendant groups are shown below:

Polyvinyl alcohol (PVA)

Polyacrylic acid (PAA)

Polyvinyl chloride (PVC)

Polystyrene (PS)

Polypropylene (PP)

Polyacrylonitrile (PAN)

1.5 Cross-linked Polymers

In addition to having linear and branched structures, polymer chains may be cross-linked; that is, they may be present as a continuous network of covalently bonded atoms, as depicted by the simulated structure in Fig. 1–4.

Traditionally, linear and branched polymers have been called thermoplastics, and cross-linked polymers have been called thermosets. These terms may be used loosely, providing one visualizes linear, branched, and network structures and recognizes that thermosetting is just one of the many ways in which a polymer may become cross-linked.

1.6 Head-to-Tail Configurations

The term "configuration" is used to describe the polymer structure that depends on the position of the covalent bonds. This term should not be confused with

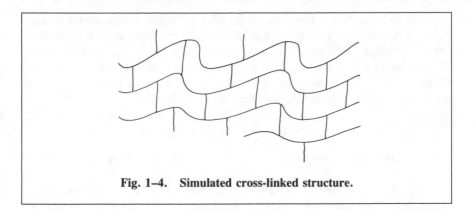

Fig. 1–4. Simulated cross-linked structure.

the term "conformation," which is used to describe various shapes of polymers resulting from "free rotation" about the single carbon-carbon bonds. The shape or conformation of a polymer chain may change readily, but changes in configurations may occur only by the cleavage of covalent bonds.

Since the repeating units in polyethylene are identical, the order in which these units are joined is inconsequential. This is not true, however, when pendant groups are present on the repeating units, as is the case with polyvinyl alcohol. If we use the customary terminology of head and tail, these repeating units could be joined as head to head or head to tail, as shown below. Fortunately, nature has simplified this problem by being biased in favor of the head-to-tail configuration.

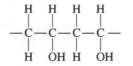
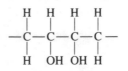

Head-to-tail configuration (PVA)   **Head-to-head configuration (PVA)**

1.7 Copolymers

Polymers consisting of multiples of the same repeating units, as in polyethylene, are called homopolymers. When pendant groups are present, as in polyvinyl alcohol, the structures may be more complicated because of the possibility of either head-to-tail or head-to-head configurations. The structure becomes even more complicated when more than one repeating unit is present in the polymer chain. Thus, one could visualize a polymer in which both ethylene and vinyl alcohol repeating units were present in the chain. For the purpose of simplicity, the different monomer repeating units will be labeled simply A and B.

So, as illustrated below, one could have a random or alternating arrangement of A and B repeating units in which the properties would be different from each other and also different from a mixture of the two homopolymers. Many commercial copolymers, such as the copolymer of ethylene and vinyl alcohol (EVA), are random copolymers. A few, such as the copolymers of styrene and maleic anhydride, are alternating copolymers.

$$-A-A-B-A-B-B- \qquad\qquad -A-B-A-B-A-B-$$

Random copolymer   **Alternating copolymer**

Copolymers may also consist of sequences of the repeating units in the chain, and these are called block copolymers. When these sequences are present as branches, the copolymer is said to be a graft copolymer. The commercial copolymer of styrene and butadiene (Kraton) is a block copolymer, and that present in the terpolymer of acrylonitrile, butadiene, and styrene (ABS) is usually a graft copolymer. These configurations are illustrated below.

$$-A-A-A-A-B-B-B-B-$$

Block copolymer

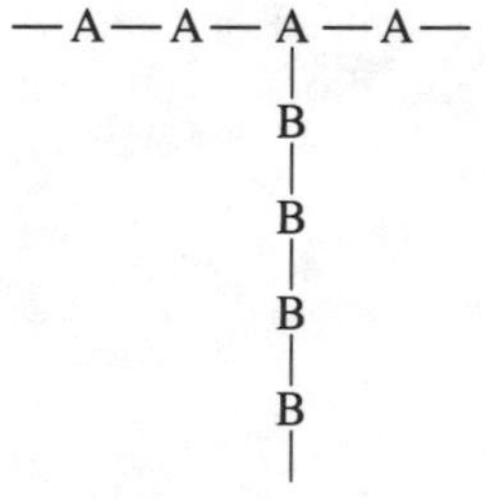

$$
\begin{array}{c}
-A-A-\ A-A- \\
| \\
B \\
| \\
B \\
| \\
B \\
| \\
B \\
|
\end{array}
$$

Graft copolymer

1.8 Tacticity

In addition to their characteristic head-to-tail configuration, homopolymers may also have specific arrangements of the pendant groups in space. In these stereospecific configurations, the pendant groups may all be on the same side of the polymer backbone, as in isotactic polypropylene (itPP); may be arranged alternately, as in syndiotactic polypropylene (synPP); or may have a random arrangement in space, as in atactic polypropylene (atPP). The most widely used commercial polypropylene is isotactic polypropylene. The different stereospecific arrangements of polypropylene are illustrated below.

```
     H   CH₃ H   CH₃ H   CH₃
     |    |   |    |   |    |
 — C — C — C — C — C — C —
     |    |   |    |   |    |
     H    H   H    H   H    H
```

Isotactic polypropylene

```
     H    H   H   CH₃ H    H
     |    |   |    |   |    |
 — C — C — C — C — C — C —
     |    |   |    |   |    |
     H  CH₃  H    H   H   CH₃
```

Syndiotactic polypropylene

```
     H    H   H    H   H   CH₃
     |    |   |    |   |    |
 — C — C — C — C — C — C —
     |    |   |    |   |    |
     H  CH₃  H   CH₃ H    H
```

Atactic polypropylene

1.9 Crystalline vs. Amorphous Structures

Structurally regular small molecules, such as sucrose, and large molecules, such as isotactic polypropylene, may form crystals. Crystallinity, of course, exists in metals, and many of the concepts of crystallinity in metals and polymers are similar.

A polymer, such as isotactic polypropylene, tends to form crystals with lamellar structures. Though it is possible to produce single crystals, the commercial polymers are not 100% crystalline. The spherulitic structure of a polymer, such as isotactic polypropylene, has amorphous components associated with folded chain surfaces. Commercial polymers, such as isotactic polypropylene, are actually semicrystalline, but the term "crystalline" is commonly used to describe a polymer possessing both crystalline and amorphous regions in the same macromolecule. Crystallinity in polymers causes opacity. Hence, films of polyethylene and polypropylene are quenched as the molten state cools in order to reduce the size and amount of crystallinity, with the goal of producing films with greater transparency.

In contrast, crystallinity is enhanced when the melt is cooled slowly and when the polymer chains are aligned during fabrication or drawing of fibers or films. As is the case for most materials, with the exception of water, bismuth, and thallium, the density of the crystalline state is greater than that of the amorphous state, and the difference between the density of the commercial polymer and that of the amorphous and crystalline polymer can be used to estimate the extent of crystallinity in the polymer. Many bulk properties of polymers, such as modulus, tensile strength, solubility, and hardness, are affected by the degree of crystallinity in the polymer.

1.10 Molecular Weight

The molecular weight of one single macromolecule is equal to the molecular weight (m) of the repeating unit multiplied by the number of repeating units (n) in the molecule. Thus, polyethylene $H(CH_2CH_2)_n H$ with 1000 repeating units would have a molecular weight of 28,002. Some natural polymers, such as specific proteins, consist of molecules with identical molecular weights and are said to be monodisperse. However, most synthetic polymers and many natural polymers, such as cellulose and rubber, consist of molecules with many different molecular weights and are said to be polydisperse. Accordingly, it is customary to use the term "average molecular weight" ($\overline{M}$) when describing the molecular weight of commercial polymers.

The molecular weight may be expressed as a number-average molecular weight ($\overline{M}_n$) or a weight-average molecular weight ($\overline{M}_w$). Obviously, $\overline{M}_n$ and $\overline{M}_w$ are identical for a monodisperse polymer, and the ratio of $\overline{M}_w$ to $\overline{M}_n$ is called the polydispersity index. This value is one for monodisperse polymers and always greater than one for polydisperse polymers.

The number-average molecular weight ($\overline{M}_n$) is simply an arithmetic mean or first moment; that is, it is the quotient when the total weight of the sample (W) is divided by the sum of the number of molecules present (N_1). Thus, a simple mixture having three molecules with molecular weights of 200,000, 300,000, and 400,000 would have a number-average molecular weight ($\overline{M}_n$) of 300,000, as this equation shows:

$$\overline{M}_n = \frac{W}{N_1} = \frac{200 \times 10^3 + 300 \times 10^3 + 400 \times 10^3}{3} = 300 \times 10^3$$

Most thermodynamic properties of polymers, including colligative properties, are related to the number-average molecular weight.

The weight-average molecular weight ($\overline{M}_w$) is the second-moment or second-power average; that is, it is the quotient when the sum of the square of the total molecular weight of the sample ($M_1^2 N_1$) is divided by the total weight of the sample (W). Thus, samples having three molecules with molecular weights of 200,000, 300,000, and 400,000 would have a weight-average molecular weight $(\overline{M}_w)$ of 322,000 and a polydispersity index of 1.07, as shown in these equations:

$$\overline{M}_w = \frac{(200 \times 10^3)^2 + (300 \times 10^3)^2 + (400 \times 10^3)^2}{200 \times 10^3 + 300 \times 10^3 + 400 \times 10^3} = 322 \times 10^3$$

$$\frac{\overline{M}_w}{\overline{M}_n} = \frac{322 \times 10^3}{300 \times 10^3} = 1.07$$

Bulk properties associated with large deformations of polymer chains, such as viscosity and toughness, are related to weight-average molecular weight values $(\overline{M}_w)$. Number-average molecular weight values may be determined by osmometry or end group analysis. Weight-average molecular weight values $(\overline{M}_w)$ may be determined by light scattering or ultracentrifugal techniques. It is customary to use viscosity (η) measurements in the laboratory and gel permeation chromatography (GPC) in production plants to determine average molecular weights $(\overline{M}_w)$. The viscosity technique is not an absolute method, and the values for the constants k and a shown below in the **Mark-Houwink equation** must be calculated from values obtained by absolute methods such as osmometry.

$$[\eta] \;=\; k\overline{M}^a$$

**Intrinsic
viscosity**

In GPC, a solution of polymer is passed through a column packed with porous beads, and the amount of eluted polymer is determined by optical or spectrophotometric methods. The retention times for macromolecules in the columns vary inversely with their size.

1.11 Intermolecular Attraction

The dissociation energy or strength of the carbon-carbon primary bonds in macromolecules is 83 kcal/mole. If these bonds are cleaved, the macromolecule is no longer intact, and the smaller segments that are produced may not be sufficiently large to function as polymers. Fortunately, unless the polymer is severely thermally degraded, the primary bonds are seldom cleaved. However, the polymer chains are attracted to each other by secondary valence bonds or van der Waals forces, which are short-range forces with much lower strength than that of primary covalent bonds, and these can be ruptured by solvents, heat, or physical force.

Van der Waals forces are the same secondary forces that aid the liquefaction of gases, raise the boiling points of liquids, and raise the melting points of solids in a homologous series. But because these forces are cumulative, they have a greater effect on the attractiveness of large molecules. Van der Waals forces are classified in accordance with their increasing energies as London dispersion forces (2 kcal/mole), dipole-dipole interaction (2 to 6 kcal/mole), and hydrogen bonds (10 kcal/mole).

It is important to note that molecules, in which hydrogen bonds are present, also have London forces. London dispersion forces or induced dipole-dipole interactions are temporary transient forces due to instantaneous fluctuations in the electron cloud density of atoms. London forces are the principal secondary valence forces present in hydrocarbon molecules, including polymers such as polyethylene and natural rubber. These forces are independent of temperature, and since they are additive, the force for a polyethylene molecule with 1000 repeating units is 6×10^{-21} kcal, compared with that of octane (C_8H_{18}), which is 2.4×10^{-24} kcal.

The calculations for polyethylene ($H(CH_2CH_2)_{1000}H$) are shown below:

$$2 \text{ kcal/mole} = 3 \times 10^{-24} \text{ kcal/molecular interaction}$$

The molecular interaction for octane $(CH_2)_8$ is:

$$8 \times 3 \times 10^{-24} = 2.4 \times 10^{-23} \text{ kcal}$$

The molecular interaction for polyethylene with 1000 repeating units is:

$$2000 \times 3 \times 10^{-24} = 6 \times 10^{-21} \text{ kcal}$$

Dipole-dipole interactions that are present, in addition to London dispersion forces in polar molecules such as polyvinyl chloride (PVC), depend on the electrostatic attractions of the chlorine atoms (Cl) in one molecule to the hydrogen atoms (H) in another molecule. These forces are temperature-dependent and may be as strong as 6 kcal/mole. They are also cumulative, and the total force in any molecule can be calculated in the same manner as that shown above for the calculation of London forces.

Very strongly polar small molecules, such as ethanol, and polymers, such as polyvinyl alcohol and cellulose, are attracted to each other by hydrogen bonds in which the hydrogen (H) atoms in one molecule are attracted to oxygen (O) atoms in another chain. Most fibers depend on hydrogen bonds for their high strength.

1.12 Chain Entanglement

In addition to depending on the intermolecular forces, which contribute to the physical properties of many types of molecules, the properties of long chain molecules, such as those present in polymers, also depend on entanglement of the polymer chains. A characteristic or critical chain length (n_c), which can be expressed as the degree of polymerization (DP), is essential before entanglement takes place.

Thus, while paraffin wax is a high-molecular-weight hydrocarbon with dispersion forces, its chain length is insufficiently long for entanglement and hence it is not a polymer. The tendency toward entanglement is related inversely to the rigidity of the chain. Thus, the critical chain length of polyethylene is much higher than that of polyvinyl alcohol.

The melt viscosity (η) of a polymer below n_c is proportional to its molecular weight. When the molecular weight exceeds n_c, the proportionality is equal to the 3.4th power of the molecular weight. Consequently, the minimum chain length required for entanglement can be determined by noting the abrupt change in viscosity when the molecular weight exceeds n_c. This relationship of melt viscosity (η) to molecular weight is shown by the following equation:

$$\eta = k\overline{M}_w^{3.4}$$

1.13 Elastomers vs. Plastics vs. Fibers

Polymers may be classified as elastomers (rubbers), plastics, or fibers in accordance with the strength of the intermolecular forces. However, there is considerable overlap in this classification. Elastomer polymers, such as natural rubber (polyisoprene), will act as plastics at very low temperatures, and fibers, such as nylon, can be used as molded plastics.

Rigid plastics and fibers are characterized by high moduli, whereas elastomers undergo large reversible elongations under applied stress. When stress is plotted against strain, the slope for fibers and rigid plastics will have low values; that is, they have a high modulus. In contrast, the elastomers will undergo reversible elongation when stretched. Prior to the yield point, a flexible plastic will exhibit a relatively high modulus, but the stress-strain curve will exhibit a plateau above the yield point, as shown in Fig. 1–5.

1.14 Polymer Synthesis by Chain-Reaction Polymerization

It is not of vital importance for design engineers and consumers to know how polymers are made. But we will present some information on these syntheses in order to make this introductory chapter more complete.

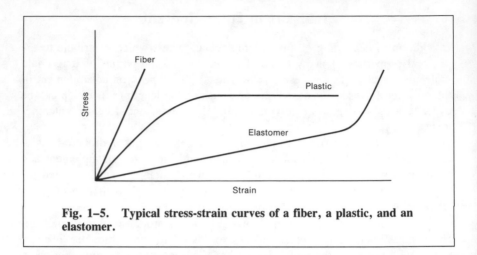

Fig. 1–5. Typical stress-strain curves of a fiber, a plastic, and an elastomer.

Natural polymers, such as cellulose, silk and wool fibers, gutta-percha plastics, and natural rubber (*Hevea braziliensis*), are supplied to the user as macromolecules. Cellulose in wood and vulcanized rubber are useful polymers but do not serve as high-performance polymers.

Synthetic polymers are produced by polymerization techniques, namely chain or addition polymerization and step or condensation polymerization. These chain reactions, which are similar to nonchemical chain reactions, will be described first. However, most high-performance polymers are produced by the condensation of difunctional reactants.

The most widely studied chain polymerization reaction is free-radical polymerization, in which an electron-deficient molecule or free radical (R.) adds to a vinyl monomer, such as ethylene, in the initiation step. As shown below, the free radical (R.) may be produced by irradiation of molecules or by cleavage of weak bonds, such as those in peroxy compounds like benzoyl peroxide.

$$C_6H_5COOCC_6H_5 \xrightarrow[\text{heat}]{\Delta} 2C_6H_5CO\cdot$$

Benzoyl peroxide

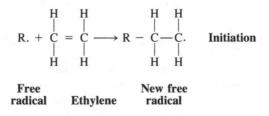

      Initiation

Free radical   **Ethylene**   **New free radical**

The newly formed free radical, which for reasons of simplicity will also be designated as R., adds to another monomer molecule (M) to produce a larger free radical, and this propagation is repeated successively to produce large macroradicals rapidly in exothermic reactions, as shown below.

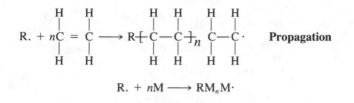

$$R. + nM \longrightarrow RM_nM\cdot$$

The polymerization reaction will be terminated when the macroradical abstracts a hydrogen atom from an unstable additive, called a chain transfer agent, when two macroradicals combine or couple, or when a hydrogen atom is abstracted by one macroradical from another macroradical to produce a saturated and an unsaturated polymer in a disproportionation step, as shown below.

$$RM_nM\cdot + HSR \longrightarrow RM_nMH + \cdot SR \qquad \textbf{Chain transfer}$$

Alkyl mercaptan   **Dead**   **New**
chain transfer agent   **polymer**   **free**
                                       radical

$$2RM_nM\cdot \longrightarrow RM_{2n+2}R \qquad \textbf{Coupling}$$

Dead
polymer

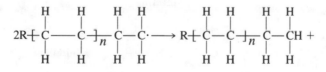

Saturated polymer

Disproportionation

Unsaturated polymer

Polymer chain reactions may also be initiated by cationic (M^+), anionic ($A:^-$), or coordination catalysts. The mechanism for the latter is complicated and is related to anionic initiation. Cations, such as the proton (H^+), may add to monomers, such as isobutylene, at very low temperatures, to produce a new cation (carbonium ion) that then undergoes propagation and chain transfer in a manner similar to that shown for free-radical polymerization. These equations are shown below.

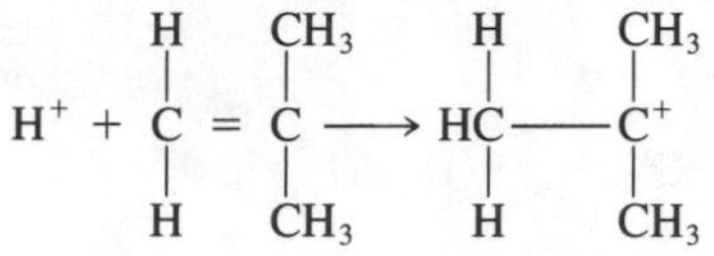

$$H^+ + \underset{\underset{H}{|}}{\overset{\overset{H}{|}}{C}} = \underset{\underset{CH_3}{|}}{\overset{\overset{CH_3}{|}}{C}} \longrightarrow HC\underset{\underset{H}{|}}{\text{——}}\underset{\underset{CH_3}{|}}{\overset{\overset{CH_3}{|}}{C^+}} \qquad \textbf{Initiation}$$

Carbonium ion

$$M^+ + nM \longrightarrow M_n M^+ \qquad \textbf{Propagation}$$

**Macrocarbonium
ion**

$$M_n M^+ + H{:}OH \longrightarrow M_{n+1}OH + H^+ \qquad \textbf{Chain transfer}$$

**Dead
polymer**

Anions, such as the butyl anion in butyllithium, may add to monomers, such as propylene, to produce a new anion (carbanion), which will then undergo propagation and chain transfer, as shown below.

$$C_4H_9Li \longrightarrow C_4H_9{:}^- + Li^+$$

Butyllithium

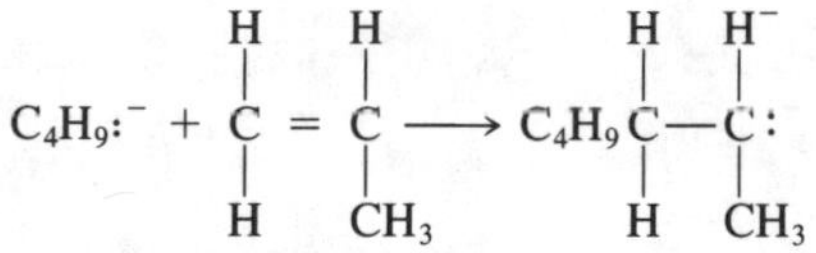

$$C_4H_9{:}^- + \underset{\underset{H}{|}}{\overset{\overset{H}{|}}{C}} = \underset{\underset{CH_3}{|}}{\overset{\overset{H}{|}}{C}} \longrightarrow C_4H_9\underset{\underset{H}{|}}{\overset{\overset{H}{|}}{C}}\text{—}\underset{\underset{CH_3}{|}}{\overset{\overset{H^-}{|}}{C}}{:} \qquad \textbf{Initiation}$$

Carbanion

$$M{:}^- + nM \longrightarrow M_n M{:}^- \qquad \textbf{Propagation}$$

$$M_n M{:}^- + H{:}H \longrightarrow M_n MH + H{:}^- \qquad \textbf{Chain transfer}$$

1.15 Polymer Synthesis by Step-Reaction Polymerization

Traditional condensation reactions involve monofunctional reactants, such as an alkyl amine (RNH_2) and a carboxylic acid ($RCOOH$). Since the products of such condensations are devoid of functional groups, they do not undergo additional condensation reactions. However, as shown below, if bifunctional reactants, such as diamines $(R(NH_2)_2)$ and dicarboxylic acids $(R(COOH)_2)$, are condensed, the products contain reactive groups that can continue to react to produce high-molecular-weight polymers.

$$RNH_2 + RCOOH \longrightarrow RCONH_2 + H_2O$$

Condensation of monofunctional reactants

$$H_2NRNH_2 + HOOCRCOOH \longrightarrow H_2NRNHOCRCOOH + H_2O$$

$$nH_2NRNHOCRCOOH \longrightarrow H_2N \overline{\left[RNHOCR \right]_n} COOH + {_n}H_2O$$

Nylon is produced by a condensation reaction. The degree of polymerization (DP) will depend on the number of step reactions that occur, and this will depend on the purity of the reactants. W. H. Carothers, the inventor of nylon, solved this problem by crystallizing the salt formed from the diamine and dicarboxylic acid before heating to cause the condensation to occur.

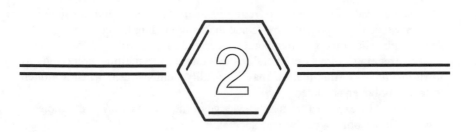

Properties of Polymers

The proper selection of polymers as materials of construction requires a knowledge of their thermal properties, such as melting point and glass transition temperature.

2.1 Melting Point of Polymers

Crystalline polymers, like metals and other crystalline compounds, undergo a phase change from solid to liquid at their characteristic melting points (T_m). This point, called the first-phase transition, is the temperature at which the crystalline solid and liquid are in equilibrium. Because of the time required for orientation of polymer chains, the melting point of polymers usually occurs over a wider range than that of metals and small organic or inorganic molecules.

Since the change in free energy (ΔG) is zero during the equilibrium melting process, the melting point (T_m) is equal to the change in enthalpy (ΔH_m) divided by the change in entropy (ΔS_m). ΔS_m is a measure of disorder and ΔH_m is a measure of the strength of the bonds holding the crystals together. The derivation of the relationship of T_m to $H_m/\Delta S_m$ is shown below.

$$\Delta G = \Delta H - T\Delta S \qquad \textbf{Gibbs free-energy equation}$$

$$\Delta G = 0 \text{ at equilibrium}$$

$$\therefore \quad T_m = \Delta H_m/\Delta S_m$$

Since considerable energy is required to increase ΔS_m and considerable energy must also be supplied to overcome the force holding the crystals together, there is a plateau in the time-temperature curve during melting.

The freezing point is essentially the same as the melting point and, as shown in Fig. 2–1, this plateau may be readily demonstrated when a molten crystalline polymer is cooled.

As might be expected from a knowledge of the increase in T_m with molecular weight in a homologous series, such as in the alkanes $(H(CH_2)_nH)$, T_m is a function of molecular weight. Because of fewer possible conformations, ΔS is lower for stiff polymers than for flexible ones, and hence T_m will be higher for stiff polymers. Obviously, the highest value of T_m will be for high-molecular-weight stiff polymers.

2.2 Glass Transition Temperature

The only thermal transition for a pure crystalline compound is the first-order transition (T_m). However, so-called crystalline polymers are only semicrystalline and contain amorphous regions. A transition, called the glass transition, also occurs in the amorphous region at a characteristic glass transition temperature (T_g). As might be anticipated from the name, this transition is readily illustrated when glass is heated.

The glass transition temperature (T_g) is the temperature at which rigid amorphous glass-like materials become flexible while the temperature is increased. Segmental motion in polymers starts at T_g and stops at this temperature when the molten polymer is cooled. There is at T_g an abrupt change in modulus, index of refraction, density, and specific heat.

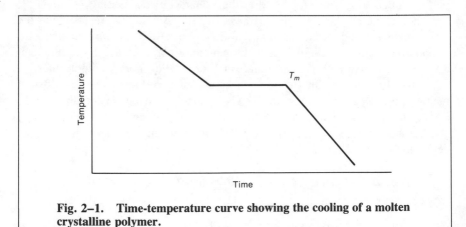

Fig. 2–1. Time-temperature curve showing the cooling of a molten crystalline polymer.

T_m values of polymers are 33% to 100% higher than T_g values on the Kelvin temperature scale. Greater differences between T_m and T_g are exhibited in symmetrical polymers, such as linear polyethylene (HDPE). T_g may be determined by measuring the change in slope when characteristic values, such as density or modulus, are plotted against temperature.

Elastomers have little elasticity below T_g and hence their useful temperature range is above that temperature. Amorphous plastics tend to creep at temperatures above T_g and so should be used only at temperatures below T_g. Polymers that contain both crystalline and amorphous regions will flow less at temperatures above T_g and below T_m because of the presence of the crystalline phase.

2.3 Softening Points

Because most of the traditionally used thermoplastics are amorphous, their resistance to heat has often been reported as a softening point, which has sometimes been confused with the melting point (T_m). The melting point, by definition, is the temperature at which the crystalline and liquid phases of a polymer are in equilibrium. The softening point of an amorphous polymer is related to T_g but should be defined to prevent confusion.

Softening points may be reported as temperatures at which a loaded indenter penetrates the polymer to a specific depth. The Vicat softening point is the lowest temperature at which a flat needle with an area of 1 mm^2 under a load of 1000 g penetrates to a depth of 1 mm.

In a more significant test outlined in ASTM D648-56, the temperature at which a bar of specific dimensions deflects 0.025 cm under a specific load is recorded as the heat deflection or heat distortion temperature. The apparatus used for this test is shown in Fig. 2–2.

2.4 Thermal Conductivity

The slow rate of heat transfer through a polymer is a major asset in the use of polymers in many applications, such as electric iron handles, pot holders, and heat shields. This thermal conductivity is reported as the K factor (thermal conductivity) in accordance with ASTM test C117-63. The relationship between the K factor, time of heat flow (Q), thickness (L), area (A), and temperatures of the hot (T_1) and cold surfaces (T_2) is as follows:

$$K = \frac{QL}{A(T_2 - T_1)}$$

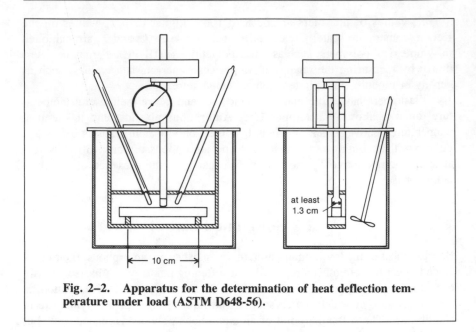

Fig. 2–2. Apparatus for the determination of heat deflection temperature under load (ASTM D648-56).

2.5 Thermal Expansion

In general, unfilled polymers expand at a rate of two to ten times that of steel. The thermal expansion of polymers is about 7 to 20 × 10^{-5}/°C, but this value can be reduced by the addition of fillers or reinforcements to the polymer. For example, the thermal coefficient of expansion of an unfilled epoxy resin with a value of about 5.5 × 10^{-5}/°C may be reduced to about 2.5 × 10^{-5}/°C by the incorporation of 40% fibrous glass in the polymer.

2.6 Electrical Conductivity

Most polymers are nonconductors, and hence they store electrostatic charges that attract dust and are capable of causing electromagnetic interference (EMI). This may be overcome by adding carbon filler or metal flakes or by coating the surface with metal.

2.7 Solubility of Polymers

Information on the solubility of selected polymers is important when these materials are used as coatings or adhesives or when solutions of polymers are used to produce fibers or films. The resistance to attack by solvents is equally important when polymers are used as materials of construction in the presence of solvents.

Prior to a half-century ago, the effect of solvents on polymers was determined by empirical tests. The guideline of "like dissolves like" was replaced by empirical numerical values, such as Kauri-butanol numbers and aniline points, and these evaluations of solubility have been displaced by the Hildebrand solubility parameter, which is related to the polarity of the solvent and the solute. The solubility parameter provides a scale of numerical values up to 23.4 Hildebrand units (H), which is the solubility parameter (δ) of water. Hildebrand solubility parameter values are usually adequate for predicting the solubility of nonpolar polymers in nonpolar solvents, but they must be modified for use with polar molecules by including contributions of dipole-dipole interactions and hydrogen bonding to solvency.

The first step in the solution process is swelling, which is followed by a breaking of intermolecular forces so that the polymer molecules are dissolved in the solvent. This process may be described by the **Gibbs free-energy equation:**

$$\Delta G = \Delta H - T\Delta S$$

in which ΔG is the change in free energy, ΔH is the change in enthalpy of mixing, and ΔS is the change in entropy.

The requirement for spontaneous dissolution is readily met when ΔG is negative, and it is assured when ΔH is less than $T\Delta S$. ΔH is related to the cohesive energy density (CED), which in turn is equal to $\Delta E/V$. CED is the heat of vaporization per unit volume. Hildebrand's solubility parameter (δ) is equal to the square root of CED.

The **Hildebrand solubility parameter** (δ) is readily calculated by using the value of ΔE or $\Delta H - RT$. In the following equation, δ is proportional to the density (ρ) divided by the molecular weight (M), which, of course, is equal to the volume (V) by definition:

$$\delta = \left(\frac{\rho(\Delta H - RT)}{M}\right)^{1/2}$$

Nonpolar polymers will dissolve in nonpolar solvents, providing that the difference in solubility parameter ($\Delta\delta$) is less than 1.8 H. Conversely, the greater the value of $\Delta\delta$, the more resistant is the polymer to that solvent. Since these values are additive, one can readily calculate δ values for mixtures of solvents and random copolymers. Block and graft copolymers have specific δ values for each domain.

The solubility parameter of nonpolar polymers may be calculated from **Small's relationship,**

$$\delta = \frac{D\Sigma G}{M}$$

in which ΣG is equal to the sum of Small's molar attraction constants.

Values for δ and G can be found in chemistry handbooks. As shown by the following equation, δ is related to the constant a in the van der Waals equation, to the surface tension γ, the index of refraction (n), the chain stiffness factor (M), and T_g:

$$\delta = \frac{1.2a^{1/2}}{V} = 4.1\frac{(\gamma^{1/3})}{(V)}0.43 = 3n^2 = M(T_g - 25)^{1/2}$$

The Hildebrand solubility concept may be readily applied to amorphous nonpolar polymers and to crystalline nonpolar polymers above T_m. Cross-linked polymers will swell in solvents with similar δ values but will not dissolve.

2.8 Transport Properties of Polymers

Both inhibition and selective transport of gases, vapors, and liquids are important. Elastomers used for balloons must resist permeation by the gas, and containers and films used for foods must have "barrier properties" that prevent migration of gases from and into the contents.

Henry's law, in which the permeation coefficient (P) is related to the diffusion coefficient (D) and the solubility coefficient (S), may be used to describe the sorption isotherm in elastomers:

$$P = DS$$

The value of S increases as the condensability of the gas increases.

Since the segmental motion is nonexistent below T_g, the sorption of gases at these temperatures is more complex and a linear, Henry's-law-type isotherm is not obtained. The isotherms of vapors sorbed into polymers below T_g have dual sorption shapes at low activity and change to the more predictable Flory-Huggins form at high activity.

The diffusion coefficient (D_0) is related to the activation energy (E). Empirical expressions for these relationships at temperatures below and above T_g are shown below.

$$\log D_0 = \frac{0.4(E_D - 10)}{1000}$$

Diffusion at temperatures below T_g

$$\log D_0 = \frac{0.5(E_D - 8)}{1000}$$

Diffusion at temperatures above T_g

It is known that permeation through amorphous polymer films depends on the rate of dissolution of the gas or liquid, the rate of diffusion of the gas through the polymer in accordance with the concentration gradient, and the energy of the low-molecular-weight materials on the opposite side of the polymer.

The relationship of the rate of diffusion to temperature may be shown by the **Arrhenius equation:**

$$D = D_0 e^{-E_d/RT}$$

When the diffusion coefficient (D) depends on the concentration of the gas or liquid, the process is said to be Fickian; that is, the weight of diffusate crossing a unit area per unit time (F) is proportional to the concentration gradient (dc/dx). The diffusate coefficient (D) is related inversely to solubility (S) and directly to permeability (P), as shown below.

$$F = -D \frac{dc}{dx}$$

Fick's law

$$D = \frac{76P}{S}$$

The solubility (S) and hence the permeation coefficient in crystalline polymers is reduced by the degree of crystallinity. Since permeability is an additive property, coextruded films may be designed that can separate gases selectively from a mixture of gases.

2.9 Electrical Properties of Polymers

Polymers such as polyacetylene will conduct electrical current, and this effect may be enhanced by the incorporation of dopants, such as arsenic pentafluoride

(AsF$_5$). Most polymers, however, are nonconductors of electricity and hence have been used for many years as electrical insulators. As shown below, the dielectric constant (ξ) is equal to the rate of the capacitance of the condenser (C) and its capacitance in a vacuum (C_0) when a dielectric material is placed between the plates. The dielectric constant (ξ), which is also called permittivity, increases with the polarity of the polymer and is temperature-dependent.

$$\xi = \frac{C}{C_0}$$

Specific resistance (P) of a polymer may be calculated from the following equation,

$$P = \frac{R(a)}{t}$$

in which R is the resistance in ohms, a is the area of the pellet under investigation, and t is the thickness.

The specific resistance (P) of copper is 10^{-6} ohm-cm, while that of PTFE and LDPE is about 10^{17} ohm-cm. The dielectric strength is the maximum voltage that a polymer can withstand before arc-through puncture occurs. The dielectric strength decreases with the thickness of the specimen.

Since alternating current (AC) is more frequently encountered than direct current (DC), it is important to observe that the electrical properties of a polymer vary with the frequency of the applied current. The response of the polymer to the applied current is delayed because of factors such as intermolecular interaction and presence of functional groups. Parameters such as relaxation time, power loss, dissipation factor, and power factor are related to this lag.

At low frequencies, the power losses are low because the dipole moments of the polymer are in phase with the charged electric field. However, the dipole moment orientation may not occur rapidly enough to keep the dipole in phase with the electric field when the frequencies are increased.

2.10 Optical Properties of Polymers

In applications requiring the transmission of visible light, polymers must be transparent, and conversely, when the transmission of specific electromagnetic radiation is undesirable, the polymer must be opaque to that radiation frequency.

Many of the optical properties of polymers are related to the refractive index (n), which is a measure of the ability of a polymer to refract, or bend, light as it passes through the polymer. As shown by **Snell's law,**

$$n = \frac{\sin i}{\sin r}$$

the index of refraction (n) is equal to the ratio of the sines of the angles of incidence (i) and refraction (r) of light as it passes through the polymer.

Values of the index of refraction for air and water are 1.000 and 1.333, respectively. The values for polymers at 25 °C vary from 1.35 for polytetrafluoroethylene (PTFE, Teflon) to 1.67 for polyaryl sulfone. The values are higher for crystalline polymers and depend on the temperature. Many amorphous thermoplastics such as polymethyl methacrylate (PMMA) and polystyrene (PS) are transparent, but their isotactic configurations are transparent or opaque.

The opacity of crystalline polymers may be reduced by quenching the molten polymer or by copolymerization. However, polymethyl pentene (TPX), which consists of amorphous and crystalline regions, is transparent because the indices of refraction of these two regions are similar.

Most polymers absorb electromagnetic radiation at characteristic wavelengths in the infrared region, and hence infrared spectroscopy is a useful tool for characterization of polymers. Some polymers, such as polyacetylene, are black because of a delocalization of electrons in the polymer. The latter contributes to the electrical conductivity of these polymers.

2.11 Polymers vs. Corrosives

With the exception of water-soluble and polar macromolecules, polymers are resistant to attack by corrosives. Fortunately, much information on the resistance to acids and alkalies can be obtained from the knowledge of the molecular structure of the polymer. Thus, the polyolefins, such as low- and high-density polyethylene (LDPE, HDPE), polypropylene (PP), and copolymers of ethylene and propylene (EPDM), are hydrocarbons, which, like the low-molecular-weight alkanes, are resistant to attack by alkalies and nonoxidizing acids. These polymers will also be resistant to weathering, but since, as shown below, polypropylene has many tertiary hydrogen atoms and LDPE has some tertiary hydrogen atoms, they will be more readily degraded when exposed to heat and sunlight than HDPE.

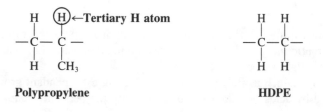

Polypropylene   HDPE

As shown below, polyvinyl chloride also has tertiary hydrogen atoms, but since it has chlorine atoms on carbon atoms adjacent to those with hydrogen atoms, it can readily undergo dehydrochlorination and produce hydrogen chloride and a dark, unsaturated polymer.

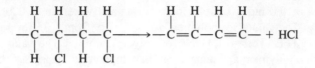

Unsaturated polymers, such as natural rubber (NR), and copolymers of butadiene or isoprene are readily attacked by ozone, chlorine, and hydrogen chloride to form ozonides, chlorinated elastomers, and rubber hydrochloride, as shown below.

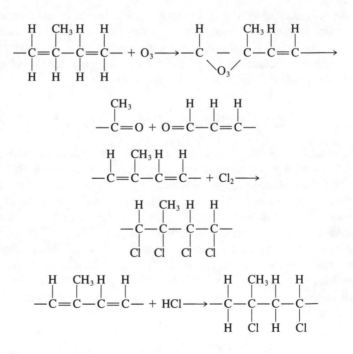

As shown above, the chlorinated and hydrochlorinated products retain their polymeric integrity but the ozonides are readily cleaved to produce lower-molecular-weight ketones and aldehydes.

Polymers, such as polyethyl acrylate, having ester pendant groups may be hydrolyzed by acids or alkalies. However, only the pendant groups are hydrolyzed and the polymer chain retains its integrity. But when the ester group is present in the polymer chain, hydrolysis causes cleavage of the polymer chains and disintegration of the molecule occurs. This is also true for polyamides, such as nylon, and for other polymers with groups not resistant to the attack by acids or alkalies. Polymers with stiffer chains, such as aromatic polyamides (aramids) and crystalline polymers, are somewhat more resistant to chemical attack. Perfluorinated polymers, such as PTFE, are extremely resistant to all types of chemical attack.

Typical reactions for functional groups may be found in organic chemistry textbooks. Resistance tables are also available in chemical handbooks. As will be discussed in Section 2.12, the weathering resistance of polymers can be improved by the addition of appropriate stabilizers.

2.12 Additives for Polymers

With the exception of undyed fibers and polymer films, most polymers are actually composites; that is, they consist of a polymer and additives. Most of the additives are functional, but in some instances nonfunctional fillers or oils are added as extenders.

2.12.1 Antioxidants

Degradation of organic polymers in the presence of oxygen and heat may be retarded by the incorporation of antioxidants, which serve as chain transfer agents and convert macroradicals to dead polymers. As stated in Section 2.10, polymers with tertiary hydrogen atoms, like polypropylene, are readily degraded when exposed to outdoor environments. This rate of degradation can be reduced dramatically, however, when small amounts of antioxidants are present.

The classical antioxidants used in natural rubber were phenyl β-naphthylamine. Hindered phenols, alkyl phosphites, and thioesters are used as antioxidants in plastics. The chain transfer of a hindered phenol and a propylene macroradical is shown in the following equation:

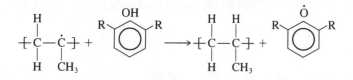

2.12.2 Ultraviolet Light Stabilizers

Ultraviolet energy from the sun's radiation is in the range of wavelengths of 280 to 400 nm, which corresponds to energy in the range of 100 to 72 kcal. This energy, which is strong enough to cleave the covalent bonds in organic polymers, will cause yellowing and embrittlement of polymers. The classical ultraviolet stabilizer, which is also used in sunscreen lotions, was phenyl salicylate, which rearranges to a 2-hydroxybenzophenone.

Hindered amines (HALS) and derivatives of 2-hydroxybenzophenones are now used widely as ultraviolet stabilizers for polymers. As shown by the following structural formula, these compounds absorb high energy, form a five- or six-membered ring (chelate), and release energy at a lower, less destructive level.

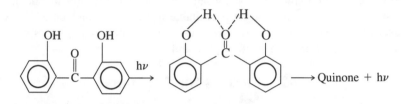

2,2′-Dihydroxybenzophenone   Chelate

2.12.3 Flame Retardants

It is now known that even flame-resistant polymers, such as PTFE (Teflon), will ignite in oxygen and that many other polymers will ignite in air. So it is essential that the flammability of materials used for construction be retarded if a fire will endanger human life or sensitive equipment.

The classical flame retardant for polymers has been a synergistic mixture of antimony oxide (Sb_2O_5) and an organic chlorine- or bromine-containing compound. Phosphates and polyphosphate esters as well as alumina trihydrate (ATH) are also used as flame retardants. Since some burning materials such as wool, nylon, and polyurethanes release hydrogen cyanide as well as carbon monoxide when burned, it is essential that the evolution of offgas from smoldering polymers be kept to a minimum.

2.12.4 Heat Stabilizers

The tendency of PVC to lose hydrogen chloride when heated may be lessened by the addition of small amounts of heat stabilizers, such as barium and cadmium salts of high-molecular-weight organic acids and epoxidized aliphatic esters.

2.12.5 Plasticizers

Because cellulose nitrate and polyvinyl chloride are intractable, their use is limited to applications in which a rigid, brittle material is acceptable. Fortunately, cellulose nitrate (often incorrectly called nitrocellulose) was flexibilized, or plasticized, by the incorporation of camphor over a century ago. Likewise, though there are many uses for rigid PVC, its use as a flexible film was made possible by the addition of alkyl esters of phthalic acid, such as dioctyl phthalate (DOP).

These additives reduce intermolecular forces and permit chain slippage and disentanglement. The rigidity of polymers may also be decreased by copolymerization. For example, the copolymer of vinyl acetate and vinyl chloride (Vinylite) is more flexible than PVC, and the copolymer of ethylene and propylene (EPDM) is a flexible elastomer in contrast to the rigid plastics HDPE and PP.

2.12.6 Impact Modifiers

The classic high-impact polystyrene (HIPS) was produced by the addition of natural rubber to polystyrene. More recently, other elastomers, such as the copolymer of butadiene and acrylonitrile (NBR), and acrylic polymers have been added to improve the impact resistance of brittle polymers.

2.13 Fillers and Reinforcements

Baekeland added wood flour to phenolic resins in order to produce a useful moldable plastic (Bakelite). Wood flour, which is made by the attrition grinding of wood waste, improves the physical properties of polymers, but many fillers, such as calcium carbonate, silica, and clay, have been used as extenders. Other, more fibrous fillers such as talc, mica, and wollastonite are functional, and their enhancement of physical properties is assured when their surfaces are treated with coupling agents.

Coupling agents, such as alkylsilanes, organotitanates, and organozirconates, are bifunctional. One of the groups in these agents is attracted to the polymer and the other is attracted to the filler surface.

The effect of the fillers on the viscosity (η) and related properties of the polymer composite can be estimated from the **Einstein-Guth-Gold (EGG) equation:**

$$\frac{\eta}{\eta_0} = 1 + 2.5C + 14.1C^2$$

The Einstein constant of 2.5 is used for spherical fillers. Other values must be used when the aspect ratio is greater than one. C is the concentration of the filler in the composite.

Fibrous fillers are more effective reinforcing agents than nonfibrous fillers. As shown by the following equation, the modulus of the composite (E_c) is equal to the sum of the moduli of the fiber (E_f) and resin matrix (E_m) multiplied by their partial volumes (V_f, $1 - V_f$) and an efficiency factor (K_{eff}):

$$E_c = k_{eff}E_fV_f + E_m(1 - V_f)$$

The most widely used fibrous reinforcement is fibrous glass. However, superior properties are obtained when the resin matrix is reinforced by aramid, graphite, boron, or ceramic fibers.

2.14 Polymer Blends

When virgin rubber was expensive, it was customary to blend natural rubber with reclaimed rubber and bitumens. More recently, LDPE has been blended with PP in order to overcome brittleness at low temperatures, and many other polymeric blends have become useful commercial plastics. The high-performance polypropylene oxide (Noryl) is simply a blend with polystyrene. The copolymer of styrene and maleic anhydride (SMA) and the terpolymer of acrylonitrile, butadiene, and styrene (ABS) are blended with several commercial polymers to provide improved properties and easier processing. Unless the components are at least marginally compatible, it is necessary to add a compatibilizing polymer, such as a block copolymer of the two polymers in the blend.

Compatibility of polymers may be estimated from their solubility parameter (δ). Good compatibility is obtained when the difference between these solubility parameters $(\Delta \delta)$ is less than 1.8 H. The thermal properties, such as glass transition temperature (T_g), of polymer blends may be a weighted compromise between those of the component polymer $(T_g 1, T_g 2)$, as shown by the **Fox equation:**

$$\frac{1}{T_g} = \frac{w_1}{T_g 1} + \frac{w_2}{T_g 2}$$

Positive synergism may exist for mechanical properties of blends such as tensile strength and modulus, but negative synergism is usually observed for impact strength of blends. The modulus E of the blend may be calculated from the following empirical equation, in which Φ is the partial value and B_{12} is an interaction parameter related to the modulus of a 50/50 blend (E_{12}), based on the empirical relationship shown below:

$$B_{12} = 4E_{12} - 2E_1 - 2E_2$$
$$E_{12} = E_1 \Phi_1 + E_2 \Phi_2 + B_{12} E_1 E_2$$

2.15 Viscoelasticity

Although there are no perfectly elastic or perfectly viscous molecules, some materials, such as metals, may be characterized as elastic materials, in which

the applied stress (S) is proportional to the strain (γ) in accordance with
Hooke's law:

$$S = G\gamma \qquad (E \cong 2.6G, \text{ below } T_g)$$

in which G is the shear modulus used in place of Young's modulus of elasticity
(E). As shown in the following equation, stress (γ) is defined by Poisson as the
ratio of longitudinal strain (γ_l) to lateral strain (γ_w):

$$\gamma = \gamma_l/\gamma_w$$

Poisson's ratio for elastomers is 0.5 but approaches 0.3 for rigid plastics,
such as PVC. It is important to note that these changes are independent of time.

Liquids, including amorphous polymers, may be characterized by
Newton's law:

$$S = \eta\frac{d\gamma}{dt} = \eta\dot{\gamma}$$

which states that the applied stress (S) is proportional to the rate of strain
($d\gamma/dt$) or applied velocity gradient. The symbol $\dot{\gamma}$ may be used in place of
$d\gamma/dt$ for the rate of strain. It is important to note that η is independent of
strain but is time-dependent. The viscosity (η) of a liquid is inversely related to
the temperature in accordance with the **Arrhenius equation,**

$$\eta = Ae^{E/RT}$$

in which E is the energy of activation for viscous flow and R is the ideal
gas constant.

Polymers possess both elastic and viscous properties and therefore are said
to be viscoelastic. Much of the initial elongation of a stressed plastic, below T_g,
is reversible and is related to the stretching and distortion of covalent bonds in
the polymer chain. Some of this elongation is also related to the disentanglement of the chains and may be reversible in the early stages. However, further
disentanglement and slippage of chains past each other is irreversible and
temperature-dependent.

In contrast to elastic materials, which return to their original dimensions
when stress is removed, there is no recovery for viscous materials, and creep
rupture will occur in viscoelastic materials if excessive stress is applied. Higher
molecular weight, which results in greater entanglement and cross-linking that
retard chain slippage, tends to retard creep at temperatures below T_g.

Processing and Fabrication of Polymers

3.1 Fibers

Silk, which is produced as a filament by the silkworm, requires only the twisting together of several filaments to produce yarn. Other natural fibers, such as cotton and wool, must be lined up by combing and drawn into continuous yarns by spinning.

Filaments of regenerated natural polymers, derivatives of natural polymers, and synthetic polymers are produced by extrusion (spinning), a process in which a melt or solution of the polymer is passed through small holes called spinnerets. Rayon filaments are obtained by passing an aqueous solution of cellulose xanthate through spinnerets and coagulating the filaments in an acid bath. Cellulose xanthate is the product of the reaction of alkali cellulose and carbon disulfide.

Acetate rayon and acrylic fibers are obtained by passing solutions of these polymers through spinnerets and evaporating the solvents. Nylon and polyurethane filaments are produced by forcing the molten polymers through spinnerets and cooling. Polypropylene filaments may be obtained by this melt spinning technique, or they may be produced by twisting and heating strips of films in a process called fibrillation. All fiber yarns may be woven on looms to produce textiles.

3.2 Castings

Film may be cast on a smooth surface from a solution or melt by using the same solidification techniques as those used for solvent or melt spinning of filaments. Thus, plasticized cellulose nitrate was cast from collodion, and the ethanol-ethyl ether solvent was evaporated to produce solvent-free celluloid films. Likewise, compositions such as beeswax, sealing wax, and bitumens were cast from hot melts of these materials. The Aztecs cast sticky, water-resistant films from natural rubber latex, and this process continues to be used to produce vulcanized rubber films. Other film-making techniques have been modified in accordance with newer developments. Thus, casting techniques have also been used to produce articles such as dentures, jewelry cases, lenses, and buttons.

Thermoset castings were obtained by Leo Baekeland in the early 1900s by the catalytic acidification of a resol liquid resin or prepolymer produced by the condensation of phenol (C_6H_5OH) and formaldehyde (CH_2O). This catalytic process, in which thermoplastic prepolymers are converted to thermosets, has been extended to urea-formaldehyde (UF), melamine-formaldehyde (MF), unsaturated polyester, and epoxy resins. UF, MF, and epoxy resins are cured by condensation reactions and the unsaturated polyesters are cured by radical chain reactions.

Acrylic sheets are cast by the radical polymerization of methyl methacrylate monomer. It is customary to heat the monomer and catalyst (initiator), such as benzoyl peroxide, until a viscous solution of the polymer, dissolved in the liquid monomer, is obtained and to continue this polymerization after the viscous liquid has been poured into a suitable mold. The mold may be two sheets of glass separated by the thickness desired for the cast sheet. The equation for this polymerization reaction is as follows:

$$n\underset{\underset{H}{|}}{\overset{\overset{H}{|}}{C}}=\underset{\underset{CH_3}{|}}{\overset{\overset{\,}{|}}{C}}-COOCH_3 \longrightarrow \left[\underset{\underset{H}{|}}{\overset{\overset{H}{|}}{C}}-\underset{\underset{CH_3}{|}}{\overset{\overset{COOCH_3}{|}}{C}}\right]_n$$

Large polyurethane (PUR) articles, such as automobile bumpers, are produced by allowing the glycol and diisocyanate reactants to react in the mold. The equation for this process, called reaction injection molding (RIM), is shown below:

$$HO-R-OH + OCNRNCO \longrightarrow \left[OR-OOC\underset{\underset{H}{|}}{\overset{\overset{H}{|}}{N}}R-\underset{\underset{\,}{}}{\overset{\overset{H}{|}}{N}}CO\right]_n$$

PVC plastisols, which consist of a suspension of finely divided PVC in a liquid plasticizer such as dioctyl phthalate (DOP), are cast by a slush molding process in which a hollow open mold is filled with the liquid plastisol and heated. The wall thickness of the casting will depend on how long the plastic is left in the mold before the unfused liquid is poured out of the mold before cooling the mold. In a similar process, called static casting, the mold cavity is filled with a finely divided powdered thermoplastic. Toys, syringes, bulbs, and containers are made by slush molding or static casting techniques.

Toys, balls, containers (including fuel tanks), luggage, and industrial parts such as arm rests are made by rotational casting. In this process, the finely divided thermoplastic is placed in a multipiece aluminum mold, which is then placed in an oven and rotated simultaneously on two axes. The rotating mold is then cooled and the mold is disassembled to obtain the fused molding.

3.3 Thermoforming

Several centuries ago, the Egyptians produced articles such as baskets and trays by placing tortoise shells or animal horns on hot forms. In the nineteenth century, piano keys were made by thermoforming celluloid sheets over wooden cores. (In fact, in spite of many advances in plastic fabrication, the name "horner" is still used to classify some workers in the plastics industry.)

Thermoplastic sheets may be thermoformed on crude male or female molds, or they may be thermoformed by much more sophisticated techniques that may include plug assists, vacuum forming, and forming with matched dies. Articles as small as blister packs and as large as automobile bodies are made by thermoforming techniques.

3.4 Foam Production

Rubber sponges have been made for many years by incorporating sodium bicarbonate into conventional compounding ingredients before vulcanizing (crosslinking) the rubber. These rubber sponges have been displaced by rubber foam, which is produced by frothing rubber latex before curing in a mold.

Expanded polystyrene (EPS, Styrofoam) is one of the most widely used thermoplastic foams. These cellular products may be obtained by extruding the polymer in the presence of a volatile compound, such as pentane.

Polymer beads, produced by the suspension polymerization process, in the presence of pentane will contain this volatile gas. These beads may be pre-expanded by heating with steam and then molded. Disposable cups and insulation boards are made from these pre-expanded beads. Of course, any thermoplastic may be foamed by the addition of a blowing agent to the molding powder. This agent may be a physical blowing agent (PBA), such as Freon, or a chemical blowing agent (CBA) produced by the thermal decomposition of a compound that decomposes to a gas when heated.

Some of the most widely used plastic foams are polyurethane (PUR) rigid and flexible foams. The gaseous propellant may be carbon dioxide, which is produced when alkyl diols and diisocyanates are reacted together in the presence of small amounts of water, or a physical blowing agent may be added. Fortunately, as evidenced from an EPS coffee cup, a skin is formed at the mold surface so that these foams may be used as structural foams. For example, the weight of ABS and PVC pipe is decreased without much decrease in utility by extruding a foam core between two thin concentric pipe extrudates.

Cellular products, in a wide range of densities, may also be produced by incorporating hollow glass or plastic beads into the polymer before molding. These syntactic foams are closed cells, but open or closed cells may be obtained by controlling the time of blowing. Open cells are produced when low-viscosity polymers are blown.

3.5 Coatings

Coatings and sealants, based on shellac and bitumens, were used 2000 years ago, and oleoresinous paints have been used for over two centuries. Pigmented cellulose nitrate coatings (DUCO) were introduced in the early 1900s. Paints, which involve the heavy metal-catalyzed polymerization of unsaturated oils in the presence of oxygen, shellac, bitumens, and cellulose nitrate, continue to be used, but they have been displaced to a large extent by waterborne, high solids, hot melt, and reactive coatings. All of these may be applied by "do-it-yourself" brush, spray, or roller coating techniques, but a high percentage of coatings is applied by more sophisticated techniques, such as extrusion, calendering, fluidized bed, electrostatic powder guns, dip, or electrodeposition techniques.

Metal wire is coated by extruding a solid or expanded plastic, such as polyethylene, onto the wire as it is pulled through the extruder die. Paperboard or other flexible sheets may be coated by squeezing the hot polymer onto the cardboard as it is passed through a series of calender rolls.

Heated metal parts, such as automobile parts, may be coated by placing them in a tank in which powdered plastics are atomized. The thickness of the coating will depend on how long the heated part remains in the fluidized bed. The coated part is then placed in an oven to ensure the fusing of the thermoplastic or to cure thermoset coatings.

In the electrostatic powder gun coating technique, a negatively charged polymer powder is sprayed on a grounded metal part. The coated articles, such as metal fences and plating racks, are then placed in an oven where the powder is fused to form a continuous coating.

Small objects, such as tool handles, may be coated by dipping the heated object in a solution or dispersion of polymer. If a plastisol or dispersion is used, the coated article should be placed in an oven in order to fuse the polymer coating.

Automobile chassis and other metal objects may be coated by placing the objects in a metal tank containing an aqueous dispersion of the polymer and by attaching a conductor to the part so that it serves as an anode. The polymer is deposited uniformly on the object when voltage is applied to the electrical circuit.

3.6 Adhesives and Laminates

With the exception that polymers must contain strong polar groups, the technology of adhesives is similar to that described for coatings in Section 3.5. When the adhesive is sandwiched in between two surfaces, the composite is called a laminate. Phenolic, urea, and melamine resin laminates have been in use for many decades. Since many of these are used in electrical applications, the National Electrical Manufacturers Association (NEMA) has developed specifications for laminates using the suffixes FR, HT, P, and so on, to designate flame retardant, high temperature, punching, and so on. NEMA grades range from X for general purpose and XX for high dielectric strength to XXX for high electrical strength, plus good mechanical properties.

Traditional phenolic laminates consisted of alternate layers of paper and resin and were available as sheets, rods, tubes, and special shapes. Cotton was used in place of paper in NEMA grades SC and CE because of its superior mechanical and electrical properties. One of the most widely used phenolic laminates (Formica, Micarta) consisted of a paper-phenolic resin laminated base covered with a colorless melamine-formaldehyde resin.

3.7 Fiber-Reinforced Plastics

Alkyd resins, produced by the condensation of trifunctional glycerol and difunctional phthalic anhydride, were used as heat-cured coatings (Glyptal) by Smith in the early 1900s. These formulations were improved by Kienle, who coined the term "alkyd" in the 1920s. Ellis and Rust increased the versatility of these polyester resins in the late 1930s by dissolving unsaturated, linear polyester prepolymers in styrene and polymerizing the mixture by free-radical techniques.

The traditional unsaturated polyesters were produced by the condensation of ethylene glycol and maleic anhydride, as shown by the following equation:

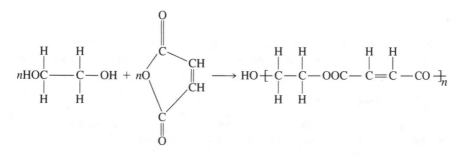

Both the ethylenic double bonds in the unsaturated polyester and those in styrene are propagated when a free radical (R·) is added to the mixture. The polymerization (curing) at ordinary temperatures is accelerated by the addition of tertiary amines, such as dimethylaniline, to the peroxy catalyst, such as benzoyl peroxide.

3.7.1 Hand Lay-Ups

In the simplest hand lay-up technique, layers of fiberglass mats, impregnated with the liquid polyester prepolymer, are placed over a suitable form and the uncured laminate is forced into place using gloved hands or a squeegee roller. It is customary to apply an unfilled polyester coat (gel coat) as the first, or surface, layer. It is also customary to restrain the laminate by use of vacuum bags and then to cure it in an oven.

3.7.2 Spray-Ups

Mixtures of chopped fibers and catalyzed polyester prepolymers may be sprayed on forms in the same manner as described for the hand lay-up techniques.

3.7.3 Filament Winding

Resin-impregnated glass filaments may be wound around a mandrel in a programmed pattern and heat cured before removing the reinforced article from the mandrel. Pipe and tanks are produced by this filament-winding technique.

3.7.4 Pultrusion

A bundle of resin-impregnating filaments may be passed through a heated die to cure the composite. This pultrusion process is used to produce pipe, profiles, and rods.

3.7.5 Molding

Mixtures of chopped glass fiber and catalyzed prepolymer may be placed in molds and heat cured. This bulk molding compound (BMC) technique is used for the production of reinforced articles. Preimpregnated sheets may also be heat formed and cured to produce bathtubs and tanks in a sheet molding compound (SMC) process.

3.7.6 Resins

The original unsaturated polyester resin has been upgraded and replaced, to some extent, by other polyesters, such as the reaction product of *bis*-phenol-A and acrylic esters, epoxy resins, and sophisticated thermoplastics.

3.7.7 Reinforcements

Fibrous glass, with appropriate silane or titanate coupling agents, continues to be used as the conventional reinforcing agent for reinforced plastics. However, aramids, graphite, and boron filaments are also used for high-performance rein-forced plastics.

3.8 Elastomer Technology

The Europeans attempted to utilize natural rubber after it was imported from Central and South America by sixteenth-century explorers. However, like the Aztecs, they obtained a sticky mass that softened in hot weather and hardened in cold weather. Fortunately, Charles Goodyear discovered vulcanization (cross-linking), a process in which rubber is heated with small amounts of sulfur at 140 °C, which overcame these disadvantages and provided a stable elastic product.

Vulcanization with sulfur is a long-time reaction that was catalyzed by the discovery of organic accelerators such as aniline and thiocarbanilide by George Oenslager in 1906. Currently, 2-mercaptobenzothiazole (Captax) and its deriva-tives are the major accelerators. The deterioration of vulcanized rubber because of aging was overcome by Moureau and Dufraisse, who in 1921 incorporated antioxidants such as phenyl β-napthylamine into rubber.

Unfilled vulcanized rubber is acceptable for rubber bands and inner tubes but lacks the abrasion resistance required for tires and mechanical goods. Fortunately, the advantages of adding large amounts of carbon black to rubber were discovered in 1910. The addition of these additives to rubber is called compounding.

The first step before compounding is mastication on a rubber mill or inten-sive mixer to reduce the molecular weight of solid rubber. The compounding ingredients are added to the masticated rubber in the order of fillers, antioxi-dants, sulfur, and accelerators. The compounded rubber may be forced through a die (extruded) or added to plies of fabric in the building up of a rubber tire. Rubber hose from an extruder or tires are then placed in an autoclave and vul-canized (cured).

Rubber latex, which may be used to make boots, is compounded by adding dispersions of the additive to the latex before dipping. Ultra-accelerators such as dialkyl thiuram disulfides, which cure at relatively low temperatures, are used as catalysts for the vulcanization of rubber latex. Rubber may also be deposited from compounded latex by electrodeposition.

Because of availability, economics, and performance, natural rubber (*Hevea braziliensis*) has been replaced in some applications by synthetic elas-tomers (SR), such as the copolymer of butadiene and styrene (SBR), the copoly-mer of butadiene and acrylonitrile (NBR), the copolymer of isobutylene and isoprene (butyl rubber), and the copolymer of ethylene and propylene (EPDM), as well as by polychloroprene (neoprene), synthetic polybutadiene, silicones, fluorocarbon elastomers, and phosphazenes.

3.9 Compression Molding

Compression molding is one of the oldest and simplest methods for converting a powdered resin into a specific shape. The mold must be cooled before the removal of the molded part when this technique is used for molding thermoplastics. Hence, with the exception of its use with high-melting thermoplastics, the use of compression molding is limited to molding thermosets.

The simple molds vary in their degree of sophistication, from simple hand molds in which an excess amount of compounded plastic is placed in the mold and heated under pressure, to positive molds in which the precise amount of molding powder is placed in the mold cavity. Excess material exudes as flash in nonpositive molds. In all cases, the molded part must be ejected after the mold is opened.

The most sophisticated type of compression molding is transfer molding, in which a measured amount of molding powder is compressed into a preform that is preheated by radiation or induction heating before being placed in the transfer pot. The temperature of the polymer is also increased further by frictional heat resulting from passage from the transfer pot to narrow chambers and orifices (runners) and gates leading to the mold cavity.

3.10 Extrusion

If one wishes to compare fabrication processes to common household operations, one might liken compression molding to the making of waffles and extrusion to meat grinding. The extruder, like the meat grinder, consists of an Archimedes' screw inside a heated cylinder (barrel). As the plastic granules, from the hopper, move along this screw, they are softened by the frictional heat and heat from the barrel.

The screw, which develops compression ratios ranging from 2-to-1 to 4-to-1, is divided into feed, compression, and metering zones. The latter zone meters the molten material before it passes through a breaker plate and enters the die. The breaker plate supports a screen pack that filters out nonpolymer contaminants and increases the back pressure.

The dimensions of extruded pipe are governed by a die mandrel and ultradie ring. The former is positioned by a spoked piece or spider and the latter is held in place by adjustable screws. The spider is eliminated when profiles are extruded. Coatings may be extruded onto continuous surfaces, such as paper board, which is then passed through a calender.

Film and sheet may be extruded through a slot die and quenched, or they may be extruded as tubes that are slit and then passed through polishing rolls. However, most film is blow extruded. The principles used in blow molding plastics are adaptations of those used in blowing glass bottles. Blow molding of containers begins with a heated parison or hollow tube that is sealed at one end and inflated by compressed air while held in a split mold and cooled before being ejected.

The blow-film process is similar to that of blowing up a rubber balloon. It begins with an extruded tube or bubble that is expanded by air until the desired thickness is obtained. The expanded tube is cooled by air from a cooling ring around the die. In some cases, the blow-extruded film is used as a seamless tubing, but it is usually slit and wound on a spool. Because of the stretching that occurs during film formation, the molecular chains are oriented and the film is stronger in the direction of orientation.

3.11 Injection Molding

The Hyatt brothers in 1872 attempted to use the injection molding process, which was a modification of die casting, but because of its explosive nature, cellulose nitrate could not be molded by this process. The emphasis during the first three decades of the twentieth century was on thermosets, and thus the interest in injection molding was not revived until the commercialization of thermoplastics in the 1930s. Thermoplastics are the most widely used plastics today, and injection molding is the principal method of fabricating both general-purpose and high-performance thermoplastics.

In the injection molding process, the plastic granules are fed from a hopper to a heated cylinder, where the polymer is heat softened and forced under pressure into a closed two-piece mold. The mold is cooled and opened, and the molded specimen is ejected. The original injection molding process used a reciprocating plunger or ram to force the molten material past a torpedo-like spreader and the cylinder wall and then to the open mold. In a more sophisticated two-stage injection molding process, the molten material is passed through a preplasticizing cylinder before going to the injection cylinder.

The preferred system is a reciprocating screw injection process, which is a modification of the extrusion process. In the injection molding process, the rotating screw acts as a ram and forces the plasticized material into the open mold cavity. The injection molding process may be used to mold articles as small as 28 g and as large as 13.5 kg or more. Modifications of injection molding can be used for molding thermosets.

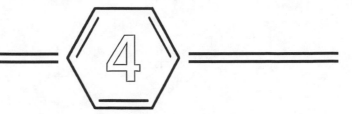

General-Purpose Polymers

Although general-purpose thermoplastics cannot be used when high performance is essential, they can be used when only moderate performance is required. Many of the thermosets are general-purpose plastics, but they may meet requirements for high performance.

4.1 Elastomers

Natural rubber (*Hevea braziliensis*) has been used for almost a century as an insulator on wire and cable and for mechanical goods. Although this product has been replaced to some extent by synthetic elastomers (SR), it still has many applications where moderate performance is acceptable.

4.1.1 Natural Rubber (NR)

Polyisoprene is produced in over 200 species of plants, including dandelions and goldenrod, but only *Hevea braziliensis* and the guayule bush (*Parthenium argentatum*) are commercially important. Rubber is synthesized in the latex of Hevea trees by the biopolymerization of isopentyl pyrophosphate. The pyrophosphate portion is eliminated during enzymatic propagation, and the product is a *cis*-polyisoprene, as shown below. In the cis isomer, the carbon-carbon bonds of the polymer are on the same side of the ethylenic double bond. In the naturally occurring, nonelastic gutta-percha and balata, the *trans*-polyisoprene is present. The carbon-carbon bonds of the polymer chain are on opposite sides of the ethylenic double bond in the trans isomer.

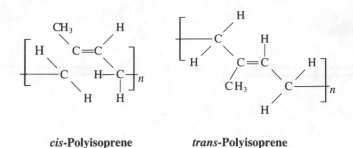

cis-Polyisoprene   *trans*-Polyisoprene

cis-Polyisoprene is polydisperse with molecular weights in the range of 100,000 to four million. After coagulation of the rubber latex, the solid rubber is masticated on a rubber mill or in an intensive mixer, such as a Banbury mixer, and the compounding ingredients are added before thermal cross-linking (vulcanization) by means of sulfur atoms.

Vulcanization of natural rubber, using a small amount of sulfur, provides a network polymer with low cross-linked density that retains its elasticity but loses its tendency to flow and become sticky. Natural rubber has been displaced, to some extent, by synthetic elastomers (SR). Four million tons of NR and 8.5 million tons of SR are consumed annually worldwide. In the United States, annual consumption is 750,000 tons of NR and two million tons of SR.

All elastomers undergo reversible elongation that may be as much as 1000 percent. The low-entropy, fully extended conformation present in the polymer chain is only one of thousands of potential conformations. This fully extended conformation is crystalline and the relatively high strength of stretched elastomers is the result of this crystallization, which is associated with polymer chain alignments of the stretched elastomers.

When linear-stretched elastomers are released, the polymer chain returns to its favorite, high-entropy, coiled conformation. Elastomers lose their characteristic elastic properties when cooled below their glass transition temperature (T_g), which is $-70 °C$ for *Hevea braziliensis*. Polyisoprene also becomes a thermoset plastic (ebonite) with a high cross-linked density when large quantities of sulfur are used for vulcanization.

The properties of natural rubber are presented in Table 4–1.

4.1.2 *cis*-Polyisoprene (IR) and *cis*-Polybutadiene (BR)

Cis polymers of isoprene or butadiene are produced by the anionic polymerization of isoprene or butadiene in the presence of butyllithium, Ziegler catalysts (titanium trichloride–diethylaluminum chloride), or Phillips catalysts (chromic oxide/silica-alumina). These synthetic polymers have a lower index of dispersity, but their physical and chemical properties are similar to those of *Hevea braziliensis*. Approximately one million tons of *cis*-polybutadiene and *cis*-polyisoprene are consumed annually in the free world.

Table 4–1. Thermal, Physical, and Chemical Properties of a Typical Vulcanized Natural Rubber(a)

Heat deflection temperature at
  445 kPa, °C..50
Maximum resistance to continuous heat, °C.......60
Coefficient of linear expansion,
  cm/cm/°C × 10^{-5} 13
Glass transition temperature, °C .... −70
Tensile strength, kPa.............................5000
Elongation, % 1000
Hardness, Shore A 80
Specific gravity 0.94
Dielectric constant................................. 2.4

Resistance to chemicals at 25 °C:(b)
  Nonoxidizing acids (20% H_2SO_4) .... S
  Oxidizing acids (10% HNO_3) .... U
  Aqueous salt solutions (NaCl)...................... S
  Aqueous alkalies (NaOH).......................... S
  Polar solvents (C_2H_5OH)......................... S
  Nonpolar solvents (C_6H_6) .... U
  Water .... S

(a) Conversion tables appear in Appendix. (b) S, satisfactory; Q, questionable; U, unsatisfactory.

4.1.3 Styrene-Butadiene Elastomer (SBR)

Polymers of 2,3-dimethylbutadiene (H—C=C—C=C—H) with groups H, CH_3, CH_3, H attached were produced early in the twentieth century by the polymerization of the 2,3-dimethylbuta-diene in the presence of sodium. This "methyl rubber" was used by the German and Russian armies during World War I, but its production was discontinued in the 1920s when Tschunker and Boch produced a random copolymer of butadiene (70%) and styrene (30%) by free-radical polymerization in an aqueous emulsion.

The Germans called their elastomer Buna S, and the U.S. National Rubber Reserve Corporation changed the name to Government Rubber-Styrene during World War II. This elastomer is now called styrene-butadiene rubber (SBR). Unlike *Hevea braziliensis,* SBR does not crystallize when extended, and hence the tensile strength of unfilled SBR is much lower than that of NR gum stock. Nevertheless, although carbon-black-filled SBR has a higher heat buildup during flexing than NR, it can still be used for passenger car tires. NR is also superior to SBR in its resilience and hot tear strength. However, SBR is more resistant than NR to abrasion and weathering.

The properties of SBR are shown in Table 4–2.

Table 4–2. Thermal, Physical, and Chemical Properties of a Typical Vulcanized Carbon-Black-Filled SBR(a)

Heat deflection temperature at
445 kPa, °C..50
Maximum resistance to continuous heat, °C.......60
Coefficient of linear expansion,
cm/cm/°C × 10^{-5} 20
Glass transition temperature, °C .... −52
Tensile strength, kPa..............................3000
Elongation, % 500
Hardness, Shore A 80
Specific gravity 1.15
Dielectric constant.................................. 2.4

Resistance to chemicals at 25 °C:(b)
Nonoxidizing acids (20% H_2SO_4) .... S
Oxidizing acids (10% HNO_3) .... Q
Aqueous salt solutions (NaCl)...................... S
Aqueous alkalies (NaOH) .... S
Polar solvents (C_2H_5OH)........................... S
Nonpolar solvents (C_6H_6) .... U
Water .... S

(a) Conversion tables appear in Appendix. (b) S, satisfactory; Q, questionable; U, unsatisfactory.

Almost three million tons of SBR is consumed annually in the free world. The leading U.S. producers and their annual production in thousands of tons are Goodyear (324), Firestone (190), Goodrich (135), and General (125). The structure of a typical repeating unit in SBR is shown below.

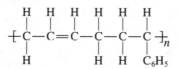

4.1.4 Butyl Rubber (IIR)

Polyisobutylene was used for many years as a flexible plastic and oil additive, but because of its high flow properties it could not be used as an elastomer. Thomas and Sparks overcame this deficiency by the cationic copolymerization of isobutylene (95%) with small amounts of isoprene (5%). These comonomers are polymerized at −100 °C by aluminum chloride ($AlCl_3$) in a methyl chloride solution.

Unlike SBR, unfilled cured butyl rubber has a high tensile strength. Because of its low degree of unsaturation, butyl rubber is characterized by excellent resistance to ozone, oxygen, and weathering. Brominated or chlorinated butyl rubber has greater resistance to permeation by gases than untreated butyl rubber and is more compatible with other synthetic rubbers than SBR. The annual production of butyl rubber (IIR) worldwide is 430,000 tons and in the United States is 135,000 tons. A typical repeating unit of this random copolymer is shown below.

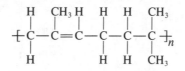

A sulfochlorinated polyethylene (Hypalon) is an amorphous elastomer-like polymer used for single-ply roof construction, covering for conveyor belts, and steam hoses and other coated goods.

4.1.5 Ethylene-Propylene Copolymer (EPPM)

High-density polyethylene (HDPE) and isotactic polypropylene (itPP) are crystalline plastics, but the random copolymer of these monomers (EPM) is a useful elastomer. EPM may be cured when a diene (D) is present with ethylene and propylene in the reactants. EPDM has a glass transition temperature of −60 °C, a brittle point of −90 °C, and a specific gravity of 0.86.

4.1.6 Polypentanomers

Linear highly unsaturated elastomers are produced by the Ziegler-catalyzed polymerization of cyclopentene, which is available as a petrochemical by-product. The glass transition temperature (−95 °C) is similar to that of amorphous cis-1,4-polybutadiene, and its crystalline melting point is just below room temperature. This elastomer is easily processed and can be cured (vulcanized) by the addition of sulfur.

4.2 Fibers

Most fibers are actually high-performance polymers, but we will discuss a few of the conventional fibers in this section.

4.2.1 Cellulose Fibers

Cotton fiber, which is obtained from the Gossypium plant of the order *Maleales*, has been used for weaving textiles for more than 3000 years. This fiber, which is almost pure cellulose, is obtained by separating the fiber from the seeds in the ginning process. As shown below by the structure for its repeating unit, cellulose is a polydisperse polymer with a $\overline{DP}$ of 3500 to 36,000, in which D-glucose is joined by beta-acetal linkages between carbon No. 1 on one unit and carbon No. 4 on the other glucose unit:

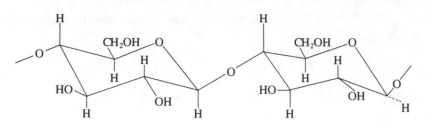

Over 14 million tons of cotton is used annually worldwide, but, because of the increased use of synthetic fibers, cotton is no longer "king."

Because of intermolecular hydrogen bonding, cellulose is insoluble in water, but alkalies will penetrate its crystal structure and produce alkali cellulose. The latter is involved in the mercerization of cotton and is used as the reactant for the production of cellulose xanthate and cellulose ether, such as ethylcellulose. The fiber in wood is cellulose.

Rayon is produced by the acidification of cellulose xanthate filaments after their extrusion through spinnerets. Acetate rayon is produced by forcing a solution of cellulose diacetate in acetone through spinnerets and then evaporating the acetone. With the exception of having longer filaments, rayon (regenerated cellulose) has properties similar to those of cotton. Cellulose acetate is a thermoplastic fiber that may be readily softened by heat and dissolved by solvents such as acetone.

Simulated formulas for cellulose xanthate and cellulose diacetate are shown below. (The repeating unit in cellulose is designated as "cell.")

$$\text{Cell} - \text{O} - \overset{\|}{\underset{S}{\text{C}}} - \overset{+}{\text{S}}, \overset{-}{\text{Na}} \qquad\qquad \text{Cell (OH) (OOCCH}_3)_2$$

Cellulose xanthate   **Cellulose diacetate**

The tensile strength of cotton is in the range of 2.7 to 4.8 g/denier, that of viscous rayon is 2 g/denier, and that of acetate rayon is 1.3 g/denier. The elongation-at-break rate of these fibers is 4 to 10%, 14 to 18%, and 22 to 28%, respectively.

Other cellulosic fibers, such as abaca (*Manila hemp*), hemp (*Cannabis saliva*), jute (*Corchorus capsularis*), and henequen are used as cordage. Jute has been used for making bags but has been displaced by polypropylene. Linen from flax, ramie from *Boehmeria nivea,* and kenaf (*Hibiscus cannabinus*) are also cellulosic fibers used interchangeably with some of the other noncotton cellulosic fibers.

4.2.2 Proteinaceous Fibers

Silk (*Bombyx mori*), which has been used for several thousand years, is a continuous filament made up of 17 different amino acids as repeating units. Despite its high cost, silk continues to be used where smooth, fine filaments are desirable. The general formula for an amino acid is shown below.

4.2.3 Mammalian Fibers

Mammalian fibers have been used for centuries as rope and textiles. These fibers, with high water-absorptive properties, consist of keratin, which is made up of repeating units of 20 amino acids. The alpha-helical structure of keratin unfolds reversibly to produce the so-called beta conformation. The beta-keratin present in feathers lacks this characteristic extensibility.

Wool, the most important animal fiber, is produced worldwide at a rate of 2.8 million tons per year. The tensile strength of wool is 1.2 g/denier and its elongation-at-break rate is 20 to 25%.

4.3 Coatings

While most coatings are not high-performance products, they do ensure the performance level of wood and steel. Oleoresinous paints, which depend on the catalyzed oxidative polymerization of unsaturated oils for film formation, have been used for many centuries for decoration and utility. Modern coatings are classified as architectural coatings, commercial finishes, and industrial coatings.

Industrial coatings are subclassified as corrosion-resistant, high-temperature, and immersion-service coatings. Pigmented cellulose nitrate (DUCO) coatings, used on automobile bodies, meet many of the requirements for high-performance materials.

Most other thermoplastic and thermoset polymers can be used as coatings, providing they adhere to the surface to be protected or to a primed surface. Alkyd and acrylic coatings are among the most widely used coatings. Both are actually polyesters, but while the ester group in alkyds is in the polymer chain, the ester group in acrylic polymers is a pendant group. About 50,000 tons of acrylic esters and 100,000 tons of alkyds are used annually as coatings in the United States.

Amorphous polyolefins, such as polyisobutylene, atactic polypropylene, and copolymers of ethylene and propylene, are applied as solution coatings; but crystalline polyolefins, such as low-density, high-density, and linear low-density polyethylene and isotactic polypropylene, must be applied as melt coatings. Polyvinyl acetate latices have been used as waterborne coatings but have been displaced, to some extent, by latices of copolymers of vinyl acetate and butyl acrylate.

The copolymer of vinyl chloride and vinyl acetate (Vinylite) continues to be used as a solution coating. The adhesion to metals of these copolymer coatings has been enhanced by the incorporation of maleic anhydride with the other monomers before polymerization.

The high degree of crystallinity in polyvinylidene chloride has been reduced by copolymerization with vinyl chloride or acrylonitrile (Saran). These copolymers have low permeability to gases and are used as barrier coatings.

4.4 Thermosets

Because they are high-temperature-resistant, general-purpose thermosets may be considered high-performance polymers. However, we will discuss some of them in this section.

4.4.1 Phenolic Resins (PF)

Phenolic resins (PF), which are products of the condensation of trifunctional phenol and difunctional formaldehyde, are widely used as adhesives for laminates and as wood-flour-filled phenolic plastics. These polymers were first commercialized by Leo Baekeland in the early 1900s. The resins used for adhesives are usually produced by condensation of the reactants under alkaline conditions. These resol resins contain adequate amounts of formaldehyde and will cure (cross-link) when heated.

PF resins used for molding compounds are usually produced by condensation of the reactants under acid conditions. These novolac PF resins are linear because they are deficient in formaldehyde. Hexamethylenetetramine, which is added to the resin before curing, serves as a source of formaldehyde. The formula for hexamethylenetetramine and the equation for the condensation of phenol and formaldehyde are shown below.

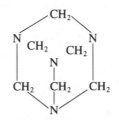

Hexamethylenetetramine (Hexa)

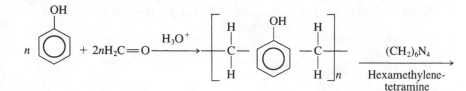

Phenol   **Formaldehyde**   **A-stage novolac**

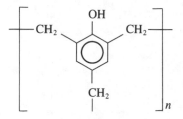

C-stage Novolac

Condensation of phenol and formaldehyde

Phenolic molding compounds are widely used in electrical and electronics applications in areas requiring good electrical properties and good physical properties at moderately high temperature. Phenolic plastics are used for electrical switch gear, circuit breakers, and wiring devices. In the United States, 1.2 million tons of phenolic resins are used annually. Of this volume, 100,000 tons is used as molding plastics, 580,000 tons as laminates, and 167,000 tons for insulation.

The properties of typical phenolic plastics are shown in Table 4–3.

4.4.2 Urea and Melamine Plastics

Since their properties are somewhat similar, urea-formaldehyde (UF) and melamine-formaldehyde (MF) resins are usually classified as amino resins. Urea formaldehyde was produced by Holzer a century ago, but it was not commercialized until 1908 when John, Goldschmidt, and Neuss and Pollack advocated its use as an adhesive and molding resin. UF molding powders have been available in the United States since 1938.

Henkel patented the condensation product of melamine and formaldehyde in 1935. Unlike the dark-colored phenolic resins, both UF and MF are colorless. Melamine resins are slightly more resistant to heat and moisture than urea resins, but as shown in Table 4–4, many of their physical properties are similar.

Both UF and MF are used as cellulose-filled molding compounds and adhesives for laminates. The molded plastics are used in wiring devices and appliance components. MF has found wide acceptance in dinnerware (Melmac). Foamed UF has been used for insulation of residential buildings, but because of

Table 4–3. Thermal, Physical, and Chemical Properties of Typical Phenolic Plastics

Property	Wood-flour-filled(a)	Mineral-filled
Heat deflection temperature at 1820 kPa, °C	165	200
Maximum resistance to continuous heat, °C	160	175
Coefficient of linear expansion, cm/cm/°C × 10^{-5}	3.0	2.0
Compressive strength, kPa	172,400	172,400
Flexural strength, kPa	62,000	82,750
Impact strength, Izod: cm · N/cm of notch	21.5	21.5
Tensile strength, kPa	48,250	41,400
Elongation, %	0.5	0.5
Hardness, Rockwell	M100	M110
Specific gravity	1.4	1.5
Resistance to chemicals at 25 °C:(b)		
Nonoxidizing acids (20% H_2SO_4)	S	S
Oxidizing acids (10% HNO_3)	Q	Q
Aqueous salt solutions (NaCl)	S	S
Aqueous alkalies (NaOH)	U	U
Polar solvents (C_2H_5OH)	S	S
Nonpolar solvents (C_6H_6)	Q	Q
Water	S	S

(a) Conversion tables appear in Appendix. (b) S, satisfactory; Q, questionable; U, unsatisfactory.

delayed release of formaldehyde its use has been discontinued in this country. Twenty thousand tons of UF molding plastics and 10,000 tons of MF molding plastics are used annually in the United States. The total annual consumption worldwide of UF and MF resins is 650,000 tons.

The structural formulas of typical repeating units of UF and MF in the thermoset polymers are shown below.

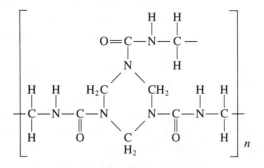

Urea formaldehyde (UF)

Table 4–4. Thermal, Physical, and Chemical Properties of Typical Amino Plastics

Property	Cellulose-filled MF(a)	Cellulose-filled UF
Heat deflection temperature at 1820 kPa, °C	150	130
Maximum resistance to continuous heat, °C	100	75
Coefficient of linear expansion, cm/cm/°C $\times 10^{-5}$	4	3
Compressive strength, kPa	276,000	220,648
Flexural strength, kPa	86,000	96,500
Impact strength, Izod: cm · N/cm of notch	16	16
Tensile strength, kPa	68,950	55,160
Elongation, %	0.7	0.7
Hardness, Rockwell	M115	M110
Specific gravity	1.5	1.5
Resistance to chemicals at 25 °C:(b)		
Nonoxidizing acids (20% H_2SO_4)	S	S
Oxidizing acids (10% HNO_3)	U	U
Aqueous salt solutions (NaCl)	S	S
Aqueous alkalies (NaOH)	S	S
Polar solvents (C_2H_5OH)	S	S
Nonpolar solvents (C_6H_6)	Q	Q
Water	S	S

(a) Conversion tables appear in Appendix. (b) S, satisfactory; Q, questionable; U, unsatisfactory.

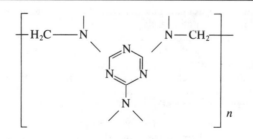

Melamine formaldehyde (MF)

4.4.3 Unsaturated Polyester Plastics

Fibrous-glass-reinforced polyesters (FRP) are used in applications requiring both moderately high and high performance. These unsaturated prepolymers, produced by the condensation of ethylene glycol and maleic anhydride, are modifications of glyptal and alkyd resin techniques. Glyptals are linear polyesters that cure when heated through esterification of the unreacted secondary hydroxyl group present in this prepolymer. Alkyds that are unsaturated cure, like oleoresinous paints, by heavy metal-catalyzed polymerization in air.

The unsaturated polyester prepolymer and styrene, which serves as a reactive solvent, cure by a free-radical mechanism. The formation of the prepolymer is illustrated by the following equation:

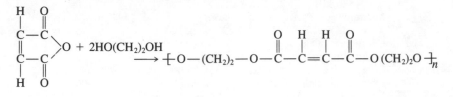

Maleic anhydride   Ethylene glycol   Unsaturated polyester prepolymer

The catalyzed prepolymer is used to impregnate fibrous glass mats or admixed with chopped fibrous glass and heat cured. The properties of a fibrous-glass-reinforced (40%) unsaturated polyester are shown in Table 4–5.

Table 4–5. Thermal, Physical, and Chemical Properties of Glass-Reinforced Unsaturated Polyesters(a)

Heat deflection temperature
  at 1820 kPa, °C....................................200
Maximum resistance to con-
  tinuous heat, °C...................................160
Coefficient of linear expansion,
  cm/cm/°C $\times 10^{-5}$ 2.5
Compressive strength, kPa..................172,000
Flexural strength, kPa 83,000
Impact strength, Izod:
  cm $\cdot$ N/cm of notch 160
Tensile strength, kPa............................69,000
Elongation, %..1.5
Hardness, Rockwell.............................. M50
Specific gravity 2
Dielectric constant.....................................5

Resistance to chemicals at 25 °C:(b)
  Nonoxidizing acids (20% H_2SO_4) .... S
  Oxidizing acids (10% HNO_3) ....Q
  Aqueous salt solutions (NaCl)...................S
  Aqueous alkalies (NaOH).......................Q
  Polar solvents (C_2H_5OH)........................S
  Nonpolar solvents (C_6H_6) ....Q
  Water ....S

(a) Conversion tables appear in Appendix. (b) S, satisfactory; Q, questionable; U, unsatisfactory.

Over 550,000 tons of fibrous-glass-reinforced polyester (FRP) is used annually in the United States, for such products as boats, storage tanks, pipe, and household appliances.

4.5 Thermoplastics

The cellulosics, namely ethylcellulose (EC), cellulose acetate (CA), and cellulose acetate butyrate (CAB), are not widely used but, as shown in Table 4–6, they possess interesting physical and chemical properties.

The "big three" general-purpose thermoplastics, namely polyolefins, polyvinyl chloride (PVC), and polystyrene (PS), account for nine million, three million, and 2.3 million tons, respectively, or almost 70% of all plastics sold annually in the United States. Some of each of these polymers is used in applications in which high performance is not required. When upgraded by copoly-

Table 4–6. Thermal, Physical, and Chemical Properties of Typical Cellulosics

Property	Ethylcellulose(a)	Cellulose triacetate	Cellulose acetate butyrate
Heat deflection temperature at 1820 kPa, °C	65	65	65
Maximum resistance to continuous heat, °C	60	60	60
Coefficient of linear expansion, cm/cm/°C $\times 10^{-5}$	15.0	12.5	14.0
Compressive strength, kPa	120,000	55,000	34,500
Flexural strength, kPa	41,370	55,200	41,350
Impact strength, Izod: cm · N/cm of notch	21.35	106.7	160
Tensile strength, kPa	34,475	41,510	34,475
Elongation, %	10	25	50
Hardness, Rockwell	R60	R80	R75
Specific gravity	1.1	1.3	1.2
Dielectric constant	3	4	4
Resistance to chemicals at 25 °C:(b)			
Nonoxidizing acids (20% H_2SO_4)	Q	U	U
Oxidizing acids (10% HNO_3)	U	U	U
Aqueous salt solutions (NaCl)	S	S	S
Aqueous alkalies (NaOH)	S	S	S
Polar solvents (C_2H_5OH)	Q	Q	Q
Nonpolar solvents (C_6H_6)	U	U	U
Water	S	S	S

(a) Conversion tables appear in Appendix. (b) S, satisfactory; Q, questionable; U, unsatisfactory.

merization, orientation, and reinforcement, however, these polymers can be used as engineering polymers.

4.5.1 Polyolefins

Prior to the late 1930s, the only commercial polyolefins available were amorphous polyisobutylene and amorphous polypropylene. The former was a soft, resinous material and the latter was a viscous liquid. Both were used as viscosity index improvers in lubricating oils. Low-density polyethylene (LDPE) was commercialized by ICI in England in the late 1930s. High-density polyethylene (HDPE) was commercialized by Phillips Petroleum and licensees of K. Ziegler in the early 1950s. Crystalline polypropylene (PP) was commercialized by licensees of Montedison and Phillips Petroleum in the early 1950s, and linear low-density polyethylene (LLDPE) was commercialized by duPont, Union Carbide, Dow, and several others in the late 1970s.

Low-Density Polyethylene (LDPE). A very small amount of low-density polyethylene (LDPE) was produced accidentally in 1933 by Fawcett and Gibson, who were attempting to react ethylene with benzaldehyde at extremely high pressures. Later, a trace of oxygen served as a free-radical initiator and produced a few grams of LDPE. The commercial product was made just before World War II by ICI using extremely high pressure in tubular reactors in the presence of a trace of oxygen.

Because of its highly branched structure, LDPE has a high volume and hence a low density. Nevertheless, it is more than 50% crystalline and, as a result, thick LDPE moldings are opaque. Over two million tons of LDPE is used annually in the United States, but this volume is being decreased. to some extent, by competition from linear low-density polyethylene (LLDPE). LDPE is widely used for blow-molded film, wire and cable coatings, and flexible tubing. The physical properties of LDPE are shown in Table 4–7.

High-Density Polyethylene (HDPE). Linear polyethylene had been made prior to 1940 by Marvel, Carothers, Mayo, and Ellis but was not commercialized until the early 1950s, when Ziegler, Zletz, and Hogan and Banks made HDPE independently. Zletz of Standard Oil of Indiana and Hogan and Banks of Phillips Petroleum produced HDPE by polymerizing ethylene in the presence of metal oxides supported on alumina, and Ziegler produced HDPE using titanium trichloride and aluminumtriethyl (Ziegler catalyst).

Because of the absence of branching, HDPE has a smaller volume and hence a higher density than LDPE. Because of this regularity, HDPE is also more crystalline and has a higher melting point than LDPE. HDPE is used as extruded pipe and extruded monofilaments, including netting, tanks, and for electrical applications. About three million tons of HDPE is used annually in

Table 4–7. Thermal, Physical, and Chemical Properties of Polyethylene

Property	LDPE(a)	HDPE	UHMWPE
Heat deflection temperature at 1820 kPa, °C	40	85	85
Maximum resistance to continuous heat, °C	40	80	80
Coefficient of linear expansion, cm/cm/°C $\times$ 10^{-5}	10.0	12.0	12.0
Compressive strength, kPa	–	20,680	–
Flexural strength, kPa	–	–	–
Impact strength, Izod: cm $\cdot$ N/cm of notch	no break	106.7	no break
Tensile strength, kPa	5,515	27,580	38,000
Elongation, %	100	30	400
Hardness, Rockwell	D40	D40	R50
Specific gravity	0.91	0.95	0.94
Dielectric constant	2.3	2.3	2.3
Resistance to chemicals at 25 °C:(b)			
Nonoxidizing acids (20% H_2SO_4)	S	S	S
Oxidizing acids (10% HNO_3)	Q	Q	Q
Aqueous salt solutions (NaCl)	S	S	S
Polar solvents (C_2H_5OH)	S	S	S
Nonpolar solvents (C_6H_6)	Q	Q	Q
Water	S	S	S

(a) Conversion tables appear in Appendix. (b) S, satisfactory; Q, questionable; U, unsatisfactory.

the United States. The physical and chemical properties of HDPE are summarized in Table 4–7.

Ultrahigh-Molecular-Weight Polyethylene (UHMWPE). While the viscosity of polymers increases rapidly, with increases in molecular weight, few desirable physical properties are improved enough to justify the increased energy required for processing, after the molecular weight exceeds a moderate value. However, improvements in stress crack resistance, tensile strength, impact strength, and stiffness of high-molecular-weight polyethylene justify the processing problem when the polymer is used as pressure pipe, extruded sheet, and large blow-molded containers. UHMWPE, with molecular weights in the range of 200,000 to 500,000 and a density greater than 0.94 g/cm^3, is produced by Ziegler or chromic oxide silica-alumina catalyzed polymerization of ethylene. The polymer must be stabilized with an antioxidant to prevent degradation during processing. The physical and chemical properties of UHMWPE are shown in Table 4–7.

Linear Low-Density Polyethylene (LLDPE). Linear low-density poly-ethylene (LLDPE) is actually a copolymer of ethylene with minor amounts of higher-molecular-weight alpha olefins, such as 1-butene or 1-hexene. The pendant groups on the linear chain increase the volume so that the density is less than that of HDPE and similar to that of LDPE. The stress crack resistance, tensile strength, and low-temperature toughness of LLDPE are superior to those of LDPE. LLDPE is displacing LDPE, to some extent, in applications such as extruded pipe, wire and cable insulation, and film. In some applications, such as blown film, it is preferable to use a blend of LLDPE with a small amount of LDPE. Over 1.3 million tons of LLDPE is used annually in the United States, and this volume is increasing.

Another linear low-density polyethylene, called very-low-density poly-ethylene (VLDPE), with density as low as 0.89 g/cm^3, is produced by using larger amounts of alpha olefin comonomers than are used for the production of LLDPE. The degree of crystallinity in VLDPE is lower than that in LLDPE or HDPE. VLDPE has a better resistance to elevated temperatures than LDPE.

Copolymers of Ethylene. As discussed in Section 4.1.5, random copolymers of ethylene and propylene (EP and EPDM) are used as ozone-resistant elastomers. In addition to EP, EPDM, and copolymers of ethylene and alpha olefins (LDPE and VLDPE), copolymers of ethylene with vinyl acetate (EVA) and with methacrylic acid (ionomers) are commercially available. EVA, produced in the United States at an annual rate of 535,000 tons, is used as film and as a melt coating. It may be hydrolyzed to produce a copolymer of ethylene and vinyl alcohol (EVAL) that is resistant to gaseous permeation.

EVA is available with 28% and 33% of vinyl alcohol. It has a tensile strength of over 7000 kPa, a density of 0.95 g/cm^3, and a ball and ring softening point ranging from 1.15 to 135 °C.

The physical and chemical properties of EVA and EVAL are shown in Table 4-8.

The sodium and zinc salts of ionomers (Surlyn) are stiff and tough plastics with excellent resistance to moderate temperatures. The physical and chemical properties of an ionomer are shown in Table 4-8.

Copolymers of ethylene and ethyl acrylate are also used to a limited extent. They have a Vicat softening point of 64 °C, a Shore A hardness of 64, a tensile strength of 50,000 kPa, and an elongation of 650%.

Polypropylene. In 1952, Hogan and Banks produced crystalline poly-propylene using a chromic oxide catalyst, supported on alumina and silica. In 1954, G. Natta used the Ziegler catalyst to produce crystalline polypropylene. After over two decades of litigation, a patent on crystalline polypropylene was granted to Hogan and Banks by the U.S. Patent Office.

Table 4–8. Thermal, Physical, and Chemical Properties of Typical Copolymers of Ethylene

Property	EVA(a)	EVAL	Ionomer
Glass transition temperature, °C	–	60	–
Heat deflection temperature at 1820 kPa, °C	–	–	45
Coefficient of linear expansion, cm/cm/°C $\times\ 10^{-5}$	22	–	15
Tensile strength, kPa	20,000	65,000	27,000
Elongation, %	500	250	400
Hardness, Shore D	30	–	60
Specific gravity	0.93	1.16	0.95
Dielectric constant	3	–	3
Resistance to chemicals at 25 °C:(b)			
Nonoxidizing acids (20% H_2SO_4)	U	U	U
Oxidizing acids (10% HNO_3)	U	U	U
Aqueous salt solutions (NaCl)	Q	S	Q
Aqueous alkalies (NaOH)	Q	Q	U
Polar solvents (C_2H_5OH)	Q	S	S
Nonpolar solvents (C_6H_6)	Q	Q	Q
Water	U	S	U

(a) Conversion tables appear in Appendix. (b) S, satisfactory; Q, questionable; U, unsatisfactory.

Atactic polypropylene is a viscous liquid or soft amorphous solid in which the methyl pendant groups are randomly placed on each side of the polymer chain. In contrast, commercial crystalline polypropylene has an isotactic arrangement in which all the methyl pendant groups are on one side of the polymer chain, as shown below.

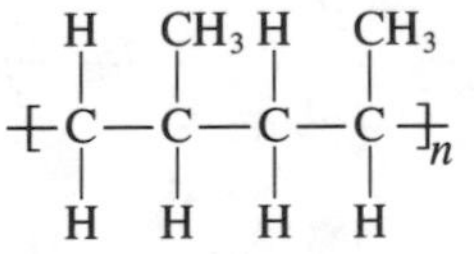

Isotactic polypropylene

Commercial polypropylene, which is highly crystalline, has a low density (0.90 g/cm^3) and a high melting point. However, it degrades when used outdoors unless an antioxidant is present. Polypropylene is used for crates, luggage, and oriented film. Over 2.3 million tons of polypropylene is used annually in the United States. Its properties are summarized in Table 4–9.

Polymethyl pentene (TPX) is an optically clear thermoplastic with a very low specific gravity (0.83). Some of its properties are shown in Table 4–9.

Polystyrene (PS). Polystyrene has been known for over 150 years but was not commercialized until the 1930s. Commercial polystyrene is an amorphous atactic brittle polymer with a high index of refraction (1.59). Because of its excellent electrical properties, it is used in many electrical applications, such as coil forms, frequency transformers, and television cabinets. It is also used for furniture and disposable dishware. Over 1.8 million tons of polystyrene is used annually in the United States. Foamed PS accounts for about 500,000 tons of PS. The formula for the repeating unit in polystyrene is shown below, and the properties of polystyrene are shown in Table 4–10.

Table 4–9. Thermal, Physical, and Chemical Properties of Typical Polypropylene and Polymethyl Pentene

Property	PP(a)	TPX
Heat deflection temperature at 1820 kPa, °C	80	55
Maximum resistance to continuous heat, °C	70	50
Coefficient of linear expansion, cm/cm/°C $\times 10^{-5}$	9.0	11.7
Compressive strength, kPa	44,800	38,000
Flexural strength, kPa	48,000	34,500
Impact strength, Izod: cm · N/cm of notch	27	27
Tensile strength, kPa	34,500	24,130
Elongation, %	100	15
Hardness, Rockwell	R80	L70
Specific gravity	0.90	0.83
Dielectric constant	2.3	2.1
Resistance to chemicals at 25 °C:(b)		
Nonoxidizing acids (20% H_2SO_4)	S	S
Oxidizing acids (10% HNO_3)	Q	Q
Aqueous salt solutions (NaCl)	S	S
Aqueous alkalies (NaOH)	S	S
Polar solvents (C_2H_5OH)	S	S
Nonpolar solvents (C_6H_6)	Q	Q
Water	S	S

(a) Conversion tables appear in Appendix. (b) S, satisfactory; Q, questionable; U, unsatisfactory.

High-Impact Polystyrene (HIPS). The inherent brittleness of polystyrene may be overcome to some extent by the incorporation of elastomers. The properties of a rubber-modified polystyrene are shown in Table 4–10.

Styrene-Acrylonitrile Copolymers (SAN). The copolymer of styrene and acrylonitrile (SAN), which is tougher and more heat resistant than PS, has

Table 4–10. Thermal, Physical, and Chemical Properties of Typical Polystyrenes

Property	Unfilled PS(a)	Impact PS	30% Glass-filled PS	SAN
Heat deflection temperature at 1820 kPa, °C	90	90	105	100
Maximum resistance to continuous heat, °C	75	70	95	85
Coefficient of linear expansion, cm/cm/°C $\times 10^{-5}$	7.5	8.0	4.0	6.0
Compressive strength, kPa	89,600	45,000	103,400	90,000
Flexural strength, kPa	82,750	50,000	117,000	100,000
Impact strength, Izod: cm · N/cm of notch	21	80	80	30
Tensile strength, kPa	41,400	41,400	82,750	60,000
Elongation, %	1.5	3	1	1.5
Hardness, Rockwell	M60	M35	M60	M80
Specific gravity	1.04	1.04	1.2	1.07
Dielectric constant	2.5	3.0	3.0	3.5
Resistance to chemicals at 25 °C:(b)				
Nonoxidizing acids (20% H_2SO_4)	S	S	S	S
Oxidizing acids (10% HNO_3)	Q	Q	Q	Q
Aqueous salt solutions (NaCl)	S	S	S	S
Aqueous alkalies (NaOH)	S	S	Q	S
Polar solvents (C_2H_5OH)	S	S	S	S
Nonpolar solvents (C_6H_6)	U	U	U	U
Water	S	S	S	S

(a) Conversion tables appear in Appendix. (b) S, satisfactory; Q, questionable; U, unsatisfactory.

been commercially available since the 1930s. About 3700 tons of SAN is used annually in the United States. The properties of this copolymer are shown in Table 4–10.

Acrylics (PMMA, PEA). As shown below, the structural formulas of the acrylates and methacrylates differ only by the presence of a methyl (CH_3) pendant group. However, these polymers differ in flexibility and resistance to hydrolysis. Polyalkyl acrylates and polyalkyl methacrylates are discussed under the heading of acrylics. Ethyl methacrylate was synthesized by Franklin and Duppa in 1865 and polymerized by Fittig and Paul in 1877. Most of the useful information required for the commercialization of acrylics was developed by Rohm in the early 1900s. These polymers were marketed under the name Plexiglas by Rohm and Haas and under the name Lucite by duPont in the 1930s.

Table 4–11. Thermal, Physical, and Chemical Properties of Typical Acrylics

Property	Cast acrylics(a)	Acrylic-PVC alloy
Heat deflection temperature at 1820 kPa, °C	95	70
Maximum resistance to continuous heat, °C	75	60
Coefficient of linear expansion, cm/cm/°C × 10^{-5}	7.0	6.0
Compressive strength, kPa	103,425	57,920
Flexural strength, kPa	96,530	72,397
Impact strength, Izod: cm · N/cm of notch	21.4	8.0
Tensile strength, kPa	65,500	44,820
Elongation, %	4	100
Hardness, Rockwell	M80	R100
Specific gravity	1.18	1.25
Dielectric constant	3.0	3.5
Resistance to chemicals at 25 °C:(b)		
Nonoxidizing acids (20% H_2SO_4)	S	S
Oxidizing acids (10% HNO_3)	U	Q
Aqueous salt solutions (NaCl)	S	S
Aqueous alkalies (NaOH)	S	S
Polar solvents (C_2H_5OH)	S	S
Nonpolar solvents (C_6H_6)	Q	Q
Water	S	S

(a) Conversion tables appear in Appendix. (b) S, satisfactory; Q, questionable; U, unsatisfactory.

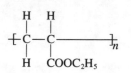

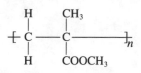

Polyethyl acrylate (PEA) **Polymethyl methacrylate (PMMA)**

The properties of typical acrylics are presented in Table 4–11, which appears at the bottom of the facing page.

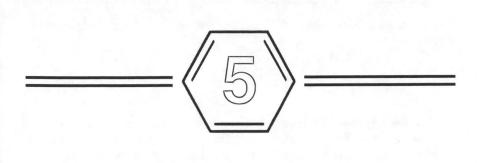

High-Performance Fibers

Many unprecedented developments have taken place in the fiber industry since the introduction of "artificial" or "mother-in-law's" silk by J. W. Swan in 1883 and Chardonnet in 1885. This cellulose nitrate filament was much too flammable to be used for clothing. Hence, it was treated with ammonium hydrosulfite to denitrate it and convert it to regenerated cellulose. Later, regenerated cellulose (rayon) was produced by the acidification of cellulose xanthate filaments, a process still in use today.

Cross and Bevan, the inventors of the viscose (xanthate) process for making rayon, produced filaments by spinning solutions of cellulose triacetate in chloroform. Because of the high costs of solvent, the cellulose triacetate was saponified to acetone-soluble cellulose diacetate, which continues to be sold as acetate rayon.

5.1 Nylon-66

The first high-performance fiber was synthesized by W. H. Carothers in the early 1930s. Carothers had already discovered that the softening point of the aliphatic polyesters, made from ethylene glycol and adipic acid, was too low for high-performance fibers. Hence, he substituted amines such as hexamethylenediamine for the ethylene glycol and produced a more desirable polyamide (melting point, 263 °C), which became known as nylon-66. Carothers recognized that high purity of his reactants was essential for obtaining high-molecular-weight polymers and developed the following relationship

(**Carothers equation**) between the degree of polymerization ($\overline{DP}$) and the extent of reaction (p):

$$\overline{DP} = \frac{1}{1 - p}$$

He was fortunate that his reactants formed a salt that could be recrystallized before it was thermally dehydrated. Thus, the value of p was so large that it was necessary to add a trace of a monofunctional acid (acetic acid) to reduce the value of $\overline{DP}$. The formula for the amine salt and for a repeating unit in nylon-66 is shown below.

$$H_2N(CH_2)_6NH_2 + HOOC(CH_2)_4COOH \longrightarrow H_2N(CH_2)_6\overset{+}{N}H_3, \overset{-}{O}OC(CH_2)_4COOH$$

Nylon-66 salt

$$nH_2N(CH_2)_6\overset{+}{N}H_3\overset{-}{O}OC(CH_2)_4COOH \xrightarrow[-nH_2O]{\Delta} \overset{}{\underset{}{+}}N(CH_2)_6\overset{H}{N}OC(CH_2)_4CO\overset{}{+}_n$$

Nylon-66

The first and second numbers after this nylon and many other available diadic nylons, such as nylon 6-10, describe the number of carbon atoms in the diamine and dicarboxylic acid, respectively.

Nylon-66 is a crystalline polymer with high strength (9 g/denier) because of the intermolecular hydrogen bonding between the hydrogen atoms on the amine groups and the oxygen atoms of the carboxyl groups. The filaments are melt spun through spinnerets and oriented by cold drawing (stretching). The drawn filaments have a density of 1.14 g/cm^3, a moderately high degree of elasticity, and a moderately high moisture absorption (4.2%).

Nylon-66 is used as a fiber for tire cord, ropes, and carpets. Nylon with more methylene groups (CH_2) in the repeating units has a lower moisture absorption. Over three million tons of nylon is produced annually in the United States. The properties of nylon are shown in Table 5–1.

5.2 Nylon-6

Gabriel and Maass obtained a hard mass by heating α, ω-aminocaproic acid in 1894, but they were unaware of any applications for this monadic polyamide. Carothers also investigated this polymer (nylon-6) in the 1930s but shelved it in favor of diadic nylons. Nevertheless, P. Schlenk produced filaments of nylon-6

Table 5–1. Thermal, Physical, and Chemical Properties of Typical Polyamides

Property	Nylon-66(a)	Nylon-6	Nylon-11	Nylon-610
Heat deflection temperature at 1820 kPa, °C	75	80	55	80
Maximum resistance to continuous heat, °C	65	70	60	75
Melting point, °C	265	215	185	220
Coefficient of linear expansion, cm/cm/°C $\times$ 10^{-5}	8.0	8.0	10	10
Tensile strength, kPa	82,750	62,000	50,000	55,000
Elongation, %	60	100	120	100
Specific gravity	1.14	1.13	1.05	1.09
Dielectric constant	4.0	4.0	3.5	4.5
Resistance to chemicals at 25 °C:(b)				
Nonoxidizing acids (20% H_2SO_4)	U	U	U	U
Oxidizing acids (10% HNO_3)	U	U	U	U
Aqueous salt solutions (NaCl)	S	S	S	S
Aqueous alkalies (NaOH)	S	S	S	S
Polar solvents (C_2H_5OH)	Q	Q	Q	Q
Nonpolar solvents (C_6H_6)	S	S	S	S
Water	S	S	S	S

(a) Conversion tables appear in Appendix. (b) S, satisfactory; Q, questionable; U, unsatisfactory.

(Perlon L) in 1937. Nylon-6 continues to be produced today under the trade names Caprolactam (Allied Chemical) in the United States, Kapron and Kaprolon in the U.S.S.R., and Amilon (Toyo) in Japan.

Nylon-6 has a lower melting point than nylon-66 but superior weather resistance. It is produced by a melt spinning process. Nylon-6 has a density of 1.13 g/cm³, a moderately high moisture absorption (4%), and a resistance to microbiological attack. Nylon-6 is used for tire cords, ropes, and textiles. The properties of nylon-6 are shown in Table 5–1, and a repeating unit is shown below.

$$\begin{matrix} H \\ | \\ {+}N(CH_2)_5\ CO{+} \end{matrix}$$

5.3 Other Nylons

Nylon-2 was produced by Leuchs in 1906 through the polymerization of anhydrocarboxyglycine, but nylon-2 polymers with sufficiently high molecular weight have not been produced commercially. Fibers of methylnylon-2 (poly-alanine) have a tenacity of 3 g/denier and an elongation of 20%.

Polymers with $\overline{DP}$ values as high as 100,000 have been produced by the low-temperature anionic polymerization of alkyl monoisocyanates in dimethyl-formamide (DMF). While the unsubstituted nylon-1 has not been synthesized, fibers have been obtained by spinning butylnylon-1 (melting point, 209 °C).

Nylon-3, produced by a molecular rearrangement and polymerization of acrylamide (H_2C=$CHCONH_2$), has been patented but is not commercially available. Dimethylnylon-3 has been produced by Farbwerke Hoechst through the polymerization of 4,4-dimethylazetidin-2-one. This high-melting polymer (melting point, 300 °C), which has been melt spun, has a moisture absorption of 4.5%.

Nylon-4, which has relatively high moisture absorption, nylon-11 (Rilsan), which has low moisture absorption, and nylon-46 (Stanyl), with a melting point of 300 °C, are all commercially available. A low-melting (185 °C), low-density (1.03 g/cm^3), high-elongation (30%) nylon (Quiana), which is produced by the condensation of *bis-p*-aminocyclohexylmethane and dodecanoic acid, is used for dress wear.

Low-melting polyamides (PA) are obtained when multiple reactants are used to produce copolymers, such as those of nylon-6 and -66 (softening point, 170 °C), nylon-6, -66, and -610 (softening point, 120 °C), and nylon-6, -66, and -12 (softening point, 120 °C). The softening points may be decreased further by the addition of *p*-toluene sulfonamide as a plasticizer.

5.4 Polyesters

As stated in Section 5.1, Carothers shelved his research on aliphatic polyesters because of their low softening points. Whinfield and Dickson upgraded polyester fibers by using terephthalic acid, which produced a stiffer polymer chain than adipic acid. Aromatic polyester fibers under the trade name Terylene were produced by ICI in England. DuPont called these polyethylene terephtha-late (PET) fibers Dacron. PET is produced by Hoechst under the trade name Trevira, by Courtalds under the trade name Lirelle, and by Eastman under the trade name Fortrel. Eastman also uses the trade name Kodel for its aromatic polyester fibers.

PET has been produced by the ester exchange reaction of dimethyl ter-ephthalate and ethylene glycol. A large *p* value in the Carothers equation is obtained by using very pure dimethyl terephthalates and distilling off the methanol produced in this exchange reaction. The equation for this reaction is shown below.

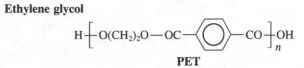

Dimethyl terephthalate   **Ethylene glycol**

PET

PET is also produced by the condensation of extremely pure terephthalic acid and ethylene glycol. It has a tenacity of 7 g/denier, a melting point of 249 °C, and a density of 1.38 g/cm^3 and is more resistant to hydrolysis than aliphatic polyesters. It is more light-stable and has a lower moisture absorption (0.4%) than nylon-66. PET, used for conveyor belts, fire hose, rope, and sailcloth, is produced at an annual rate of 5.5 million tons worldwide and is now a dominant synthetic fiber.

Kodel (melting point, 290 °C) is produced by the condensation of terephthalic acid and cyclohexanedimethanol. Another polyester produced by the condensation of terephthalic acid and *p*-xylene diol has a melting point of 253 °C. The properties of polyester fibers are shown in Table 5–2, and the formulas for the two glycols mentioned above are shown below.

HOCH$_2$—⬡—CH$_2$OH   HOCH$_2$—◯—CH$_2$OH

Cyclohexanedimethanol   **p-Xylene diol**

Polyesters may also be produced by the ring opening of lactones, such as ξ-caprolactone. The formula for polycaprolactone is as follows:

$$+(CH_2)_5 - COO +_n$$

A fiber called Fiber K has been made from polypivalactone by Shell Oil. This polymer has a high melting point (235 °C), a density of 1.8 g/cm^3, and a moisture absorption of 1.2%. Chemists at American Cyanamid have produced high-melting fibers (melting point, 230 °C) by the melt spinning of polyglycollide obtained by the cationic polymerization of hydroxyacetic acid. This fiber, called Dexion, has poor hydrolytic stability and may be used for surgical sutures.

The polyester produced by the condensation of ethylene glycol and terephthalic acid and *p*-hydroxybenzoic acid was marketed under the trade name

Table 5–2. Thermal, Physical, and Chemical Properties of a Typical Polyester(a)

Heat deflection temperature at
  1820 kPa, °C 65
Maximum resistance to continuous
  heat, °C...60
Melting point, °C 240
Coefficient of linear expansion,
  cm/cm/°C × 10^{-5} 7
Tensile strength, kPa......................... 60,000
Elongation, % 50
Specific gravity 1.35
Dielectric constant................................ 3.0

Resistance to chemicals at 25 °C:(b)
  Nonoxidizing acids (20% H_2SO_4) .... Q
  Oxidizing acids (10% HNO_3) .... Q
  Aqueous salt solutions (NaCl).................. S
  Aqueous alkalies (NaOH)...................... Q
  Polar solvents (C_2H_5OH)...................... Q
  Nonpolar solvents (C_6H_6) .... U
  Water .... S

(a) Conversion tables appear in Appendix. (b) S, satisfactory; Q, questionable; U, unsatisfactory.

Grilene. Another polyester ether (A-Tell) (melting point, 222 °C) is produced by the condensation of p-hydroxybenzoic acid and ethylene oxide. The formulas for these reactants are shown below.

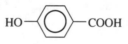

p-Hydroxybenzoic acid   Ethylene oxide

5.5 Aromatic Polyamides (Aramids)

An aromatic polyamide (nylon 6-T) is produced by the condensation of hexa-methylenediamine and terephthalic acid. This high-melting polymer (melting point, 370 °C) has a density of 1.21 g/cm³ and a moisture absorption of 4.5%. The properties of a typical aramid are shown in Table 5-3, and the equation for the production of this high-melting fiber is shown below.

**Table 5–3. Thermal, Physical, and Chemical Properties of a
Typical Aramid(a)**

Heat deflection temperature at
1820 kPa, °C 260
Maximum resistance to continuous
heat, °C 150
Coefficient of linear expansion,
cm/cm/°C $\times 10^{-5}$ 2.6
Tensile strength, kPa 120,000
Elongation, % 5
Specific gravity 1.2
Dielectric constant 3.0

Resistance to chemicals at 25 °C:(b)
Nonoxidizing acids (10% H_2SO_4) .... Q
Oxidizing acids (10% HNO_3) .... U
Aqueous salt solutions (NaCl) .... S
Aqueous alkalies (NaOH) .... S
Polar solvents (C_2H_5OH) .... Q
Nonpolar solvents (C_6H_6) .... S
Water .... S

(a) Conversion tables appear in Appendix. (b) S,
satisfactory; Q, questionable; U, unsatisfactory.

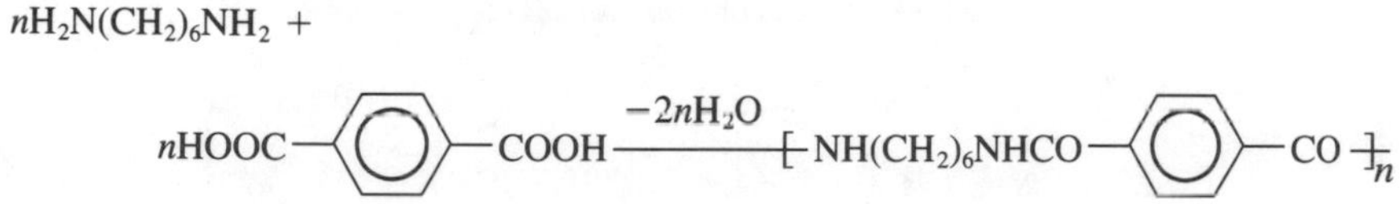

nH$_2$N(CH$_2$)$_6$NH$_2$ +

5.6 Poly p-Benzamide (Kevlar 49, Twaron)

Poly p-benzamide (Kevlar 49, Twaron) is produced by the condensation of
p-phenylenediamine and terephthaloyl chloride. This strong, high-melting fiber
is used in tires, bullet-proof vests, cables, and as a reinforcing fiber for polyester and epoxy resins. A comparable fiber (HM50) is produced by using the
poly p-benzamide reactant plus 3,4-diaminodiphenyl ether. The repeating unit
in poly p-benzamide is shown below.

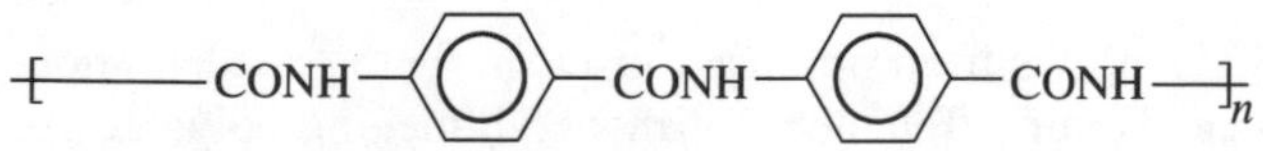

Poly m-phenylene isophthalamide (Nomex) is produced by the condensation of m-phenylenediamine and isophthaloyl chloride. This high-melting
(370 °C) fiber is dry spun from a solution of the polymer in dimethylformamide

in the presence of a small amount of lithium chloride. Nomex is used for space suits, racing drivers' suits, gas filters, dust collectors, conveyor belts, and paper-making dryer felt. The formula for the repeating unit in this high-density (1.38 g/cm^3) polymer is as follows:

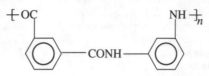

Comparable aromatic polyamides have been made by the condensation of piperazine with o-phthaloyl chloride, by the condensation of m-phenylenedi-amine with adipic acid, and by the condensation of m-phenylediamine with sebacic acid. The repeating units in these aromatic polyamides are shown below.

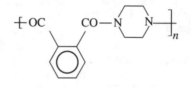

Piperazine polyamide

Nylon MXD-6

$$\text{+ N}-\text{CH}_2-\bigcirc-\text{CH}_2\text{NHCO(CH}_2)_8 \text{ CO +}_n$$

Poly m-phenylene sebacamide

5.7 Polyureas

Polyurea (Urylon) has been produced in Japan by the condensation of urea and nonamethylenediamine. Filaments of this high-melting polymer (melting point, 240 °C) are obtained by melt spinning. The properties of this low-density fiber (1.07 g/cm^3) are similar to those of nylon-66. The formula for the repeating units in polyurea is shown below.

$$\text{+ (CH}_2)_9\text{NHCONH +}_n$$

5.8 Polyurethanes

Stiff, low-melting (melting point, 175 °C) polyurethane fibers, under the trade name Perlon U, were produced by O. Bayer in the 1940s. As shown by the following equation, the pioneer polyurethane fibers were produced by the condensation of butanediol and hexamethylene diisocyanate. Filaments that were cut up and used for brush bristles and filter cloth were obtained by melt spinning of this polymer.

$$nHO(CH_2)_4OH + nOCN(CH_2)_6NCO \longrightarrow \left[O(CH_2)_4OOCNH(CH_2)_6NHCO \right]_n$$

Spandex, a more useful elastic block copolymer, was developed by duPont chemists in 1954. These "snap-back" fibers consist of a long, soft block of a polyether and a short, hard block of a polyurethane. Fibers competitive with duPont's Lycra have been made by Uniroyal under the trade name Vyrene.

The recovery of these "snap-back" fibers is inferior to that of rubber, but spandex fibers have an elongation of more than 500%. These filaments have been used in place of elastomeric fibers in swimming suits and foundation garments. Simulated structures of these unstretched and stretched filaments are shown in Fig. 5–1.

5.9 Nitrile Fibers (Acrylic Fibers)

The potential for polyacrylonitrile (PAN) fibers was recognized for many years, but PAN could not be melt spun and no good solvent for this strongly hydrogen-bonded polymer was available. After many Edisonian trials, chemists in Germany and at duPont in the United States discovered independently that filaments of PAN could be dry spun from solutions of the polymer in dimethylformamide (DMF).

DuPont chemists used the trade name Orlon to describe their strong PAN fibers, which had a tenacity of 3.5 g/denier, a moisture absorption of 0.5%,

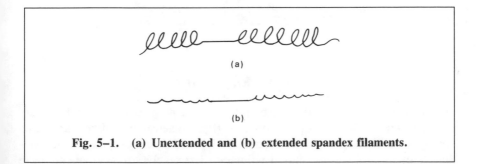

(a)

(b)

Fig. 5–1. (a) Unextended and (b) extended spandex filaments.

and a density of 1.2 g/cm^3. PAN (melting point, 250 °C) is stable at relatively high temperatures but is hydrolyzed by hot alkaline solutions. Extremely high temperature causes a loss in weight and cyclization. The cyclized product, called "black Orlon," is used as graphite fiber. The equation for the transformation of repeating units of PAN to the cyclic product is shown below.

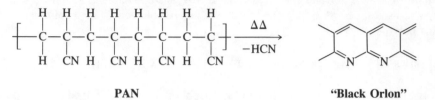

PAN   "Black Orlon"

Because of difficulties in dyeing PAN, dyeable groups were introduced by copolymerization. Dralon, produced by Farbenfabriken Bayer AG; Redon, produced in Austria; Zefran, produced by Dow; Zolan, produced by Zellwolle AG; Creslan, produced by American Cyanamid; Courtelle, produced by Courtalds; Crylor, produced by Rhodiadeta; and Acrylan, produced by Monsanto, are copolymers of acrylonitrile ($H_2C{=}CHCN$) with up to 15% of a comonomer, such as vinylpyridine.

The U.S. Federal Trade Commission has ruled that the term "acrylic" be used only with fibers containing at least 85% acrylonitrile, and that the term "modacrylic" be used for fibers produced from copolymers containing 35% to 85% acrylonitrile. Verel, produced by Eastman, and Teklan, produced by Courtalds, are modacrylic fibers.

A copolymer of vinylidene nitrile and vinyl acetate has been marketed in Europe under the trade name Darvan and in the United States under the trade name Travis. While this alternating copolymer is of interest scientifically, it is no longer available commercially. Alternating copolymers are produced regardless of the ratio of the reactants. As shown below, Darvan has a density of 1.18 g/cm^3, a moisture absorption of 2.5%, and a softening point of 175 °C.

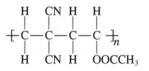

5.10 Polyolefins

Monofilaments and fibrillated fibers have been produced from high-density polyethylene (HDPE) and polypropylene (PP). Whereas most fibers are based on hydrogen-bonded macromolecules, polyolefin fibers depend on the symmetry of these linear-structured polymers. Extremely strong filaments have been obtained by drawing and orienting polyolefins in the gel state.

5.10.1 Low-Density Polyethylene (LDPE)

Relatively high-molecular-weight (21,000), low-density polyethylene (LDPE) when heated to 300 °C may be melt spun, and the filaments may be cooled and drawn to about 600% of their original length. These filaments have the characteristic low density of LDPE (0.9 g/cm³), low softening point (100 °C), low frictional coefficient, moderately high tenacity (3 g/denier), and moderate elongation (4%). LDPE fibers have been used to produce protective clothing, upholstery, and filter cloths.

5.10.2 High-Density Polyethylene (HDPE)

With the exception of density (0.96 g/cm³) and softening point (125 °C), the production of properties of high-density polyethylene (HDPE) fibers is similar to the techniques cited for LDPE. HDPE fibers are used for automobile interiors and ropes.

5.10.3 Polypropylene (PP)

Because of structural regularity, the density of isotactic polypropylene (PP) fibers is lower (0.91 g/cm³) and the melting point is higher (170 °C) than that of HDPE. PP may be melt spun, cooled, and drawn to produce filaments with 68% crystallinity. These fibers have relatively high tenacity (9 g/denier), good elongation (20%), and very low moisture absorption (0.05%). PP fibers are used for ropes, fishnets, mats for soil stabilization, carpets, and garments.

PP fibers are also produced by the fibrillation, or split film, process in which stretched film strips are twisted (fibrillated) to produce fiber. Because of the presence of a tertiary hydrogen atom in the PP chain, antioxidants must be present if the PP fiber is used out of doors. A section of an isotactic polypropylene chain is shown below.

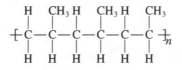

Crystalline polypropylene

5.11 Polyfluorocarbon Fiber

Filaments have been produced from polyvinyl fluoride (Tedlar), polyvinylidene fluoride (Kynar), polychlorotrifluoroethylene (Kel F), and polytetrafluoroethylene (PTFE, Teflon), but only PTFE fibers have been used to any great extent. As is the case for polyolefins, the strength of PTFE fibers depends not on hydrogen bonding but on the symmetry of the polymer molecule that favors good packing and crystallinity. The close packing is responsible for high cumulative intermolecular dispersion forces and high density (2.2 g/cm^3).

The soluble polyfluorocarbons can be readily dry spun, but the insoluble PTFE requires rapid sintering at 385 °C of a weak yarn produced by the alignment of minute polymer particles produced by extruding an aqueous colloidal dispersion through a capillary tube into an acid bath. PTFE fibers have moderately low tenacity (1.5 g/denier), moderate elongation (13%), and zero moisture absorption. They are used for pulp shaft packing, filtration fabrics, and protective clothing. Microporous PTFE fibers, which have been used for rainwear (Gore-Tex), are also used for surgical implants. The repeating unit in the PTFE polymer chain is shown below.

5.12 Polyimides

Polyimide (PI) fibers are produced by solvent spinning the soluble linear polyamic acid and heating the filaments at about 300 °C to form the intractable polyimide. The precursor, polyamic acid, is produced by the condensation of pyromellitic dianhydride and *bis*-4-aminophenyl ether. Polyimide fibers are used for electric locomotive motor insulation and other high-temperature applications. The properties of a typical polyimide are shown in Table 5–4, and the formula for the repeating unit of a polyimide is shown below.

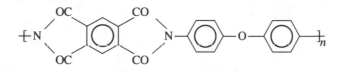

High-melting (melting point, 590 °C) polybenzimidazole (PBI) fibers are also used when resistance to high temperature is essential. PBI is used in escape suits and safety garments in space travel. The formula for the repeating unit in PBI is shown below.

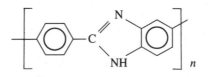

**Table 5–4. Thermal, Physical, and Chemical Properties of a Typical
Polyimide (PI)(a)**

Heat deflection temperature at
 1820 kPa, °C 315
Maximum resistance to continuous
 heat, °C...300
Coefficient of linear expansion,
 cm/cm/°C × 10^{-5} 5.0
Tensile strength, kPa...........................96,500
Elongation, % 8
Specific gravity 1.4
Dielectric constant...................................3.5

Resistance to chemicals at 25 °C:(b)
 Nonoxidizing acids (20% H_2SO_4) ....Q
 Oxidizing acids (10% HNO_3) ....Q
 Aqueous salt solutions (NaCl)...................S
 Aqueous alkalies (NaOH).......................U
 Polar solvents (C_2H_5OH)........................S
 Nonpolar solvents (C_6H_6) ....S
 Water ....S

(a) Conversion tables appear in Appendix. (b) S,
satisfactory; Q, questionable; U, unsatisfactory.

5.13 Graphite Fibers

J. W. Swan produced graphite filaments by the pyrolysis of paper board, cotton
threads, and rayon, and Edison produced these filaments in the nineteenth century by the pyrolysis of bamboo. Comparable graphite fibers with higher
modulus are now made by the pyrolysis of polyacrylonitrile (PAN) filaments.
These graphite fibers have a tensile strength of over one million kPa and a
Young's modulus of 180 million kPa, which corresponds to a tenacity of about
10 g/denier. Graphite fibers are also produced by the pyrolysis of pitch. These
fibers, which are marketed under the trade names Thornel, Hyfil, Grafil, and
Moinoc, are used as reinforcements for polyester and epoxy resins. The properties of graphite fibers are summarized in Table 5–5.

5.14 Fibrous Glass

Glass-like fibers called Pele's hair were found in the vicinity of volcanoes
many centuries ago. In 1841, a glass-fiber-making machine was invented in
England. This spinning process, in which the molten glass passed through fine
orifices (spinnerets), was upgraded by Owens-Illinois and Corning engineers in

Table 5-5. Thermal, Physical, and Chemical Properties of Typical Graphite Fibers (a)(b)

Tensile strength, kPa.................... 2,000,000
Elongation, %.. 0.6
Specific gravity 1.63
Modulus of elasticity, kPa   550,000,000

(a) Graphite fibers are resistant to most corrosives and solvents. (b) Conversion tables appear in Appendix.

1938. Owens-Corning Fiberglas continues to produce very large quantities of fibrous glass.

These high-strength, amorphous fibers (6 g/denier) have low elongation (2%) and relatively high density (2.6 g/cm^3). Glass fibers have low water absorption and good resistance to flame. They are used as filter cloths, lamp wicks, ropes, and insulation. However, the principal use of fibrous glass is as a reinforcement for polyester and epoxy resins.

5.15 Other Inorganic Fibers

A silica fiber (Sil-temp, Refrasil) is produced by the acid leaching of fibrous glass. Quartz fibers have been used as heat-resistant fibers for textiles. Fiberfrax, which is an aluminum silicate fiber, is also used as a heat-resistant and insulation material. Boron filaments are produced by the deposition of boron on a continuous filament, such as stainless steel. The boron is produced by the decomposition of boron tetrachloride in the presence of hydrogen.

About 5% by weight of stainless steel fiber has been added to polymers to provide shielding from EMI/RFI interference. Comparable results are observed when 50% of nickel-coated graphite fibers are added to polymers such as ABS. An alumina/silica fiber is currently being produced by Carborundum.

High-Performance Elastomers

Since there are no other materials with comparable elastic qualities, even natural rubber (discussed in Section 4.1.1) is actually a high-performance polymer. This same statement could also be applied to synthetic polyisoprene and polybutadiene (Section 4.1.2), SBR (Section 4.1.3), butyl rubber (Section 4.1.4), and EPDM (Section 4.1.5). The discussion in this chapter, however, will be limited to polymers whose properties are superior to those of general-purpose elastomers.

6.1 Neoprene

Not only was neoprene (CR) discovered accidentally, but its discovery was also based on the synthesis of an explosive vinylacetylene by a priest, J. A. Nieuwland, who was a professor of botany at the University of Notre Dame. Nieuwland produced vinylacetylene ($CH_2 = CH - C \equiv CH$) by passing acetylene into a concentrated solution of copper (I) and ammonium chloride. Collins and Carothers at duPont produced chloroprene in 1931 by the accidental addition of hydrogen chloride to vinylacetylene and then polymerized chloroprene to polychloroprene by free-radical chain emulsion polymerization. The polymer was called Duprene, but this name was changed to neoprene.

Neoprene is more resistant to heat, abrasion, and solvents than NR. It has a tensile strength greater than 3000 psi, a density of 1.23 g/cm^3, and a Shore durometer A hardness of 40 to 95, depending on the vulcanization formulation. The principal configuration present in the commercial product is that of *trans-*

chloro-2-butenylene, and this configuration governs the crystallinity of CR. Neoprene is used for wire, hose, cable, belts, gaskets, and molded articles.

In the United States alone, 275,000 tons of neoprene is produced annually. This elastomer is also produced in England, Germany (Perbunan), and the U.S.S.R. (Sovprene, Najrit). The properties of carbon-reinforced vulcanized neoprene are shown in Table 6–1, and the repeating unit in neoprene is shown below.

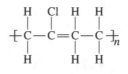

6.2 Thiokol

Like neoprene, polyethylene sulfide (Thiokol) was discovered accidentally by a physician, J. C. Patrick, who was attempting to produce ethylene glycol. Patrick tried to hydrolyze 1,2-dichloroethane by heating it with sodium polysulfide. Subsequently he used β-chloroethyl formal as the reactant to produce Thiokol FA. Thiokol FA has a density of 1.34 g/cm^3 and, depending on the compounding formulation, a Shore durometer A hardness of 35 to 80. This odoriferous elastomer was used as a solvent-resistant material in 1929.

The principal use of Thiokol is as a liquid polymer converted by the addition of lead dioxide to a solid elastomer at ordinary temperatures. The liquid

Table 6–1. Thermal, Physical, and Chemical Properties of Carbon-Reinforced Vulcanized Neoprene (CR)(a)

Glass transition temperature, °C .... −42
Tensile strength, kPa.....................27,000
Elongation, % 500
Hardness, Shore A 75
Specific gravity 1.42

Resistance to chemicals at 25 °C:(b)
   Nonoxidizing acids (20% H_2SO_4) .... S (HCl)
   Oxidizing acids (10% HNO_3) .... Q
   Aqueous salt solutions (NaCl)............ S
   Aqueous alkalies (NaOH) .... S
   Polar solvents (C_2H_5OH) .... Q
   Nonpolar solvents (C_6H_6) .... Q
   Water .... S

(a) Conversion tables appear in Appendix. (b) S, satisfactory; Q, questionable; U, unsatisfactory.

polymers (LP-2, LP-3) are produced by reducing Thiokol FA by the addition of sodium sulfide (NaSH) and sodium sulfite (Na$_2$SO$_3$). Since LP-2 and LP-3 are used as binders for solid propellants, they are truly high-performance polymers. The structural formula for the repeating unit in Thiokol is shown below, and the properties of Thiokol are shown in Table 6–2.

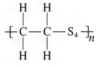

6.3 Acrylonitrile-Butadiene Copolymer Elastomers (NBR)

Butadiene and acrylonitrile were copolymerized in emulsion by Conrad and Tschunker in 1930 to produce a solvent-resistant elastomer called Buna N. The oil-resistant properties of Buna N increase and the low-temperature flexibility decreases as the acrylonitrile content is increased. Those elastomers containing a high percentage (45%) of acrylonitrile were called Perbunan. Goodrich called this copolymer Hycar, Goodyear called it Chemigum, and Firestone called it Butaprene N. It is now called NBR.

While most of today's tires are made from SBR, the first American tires were made from NBR. NBR is used for fuel-cell liners, fuel hose, oil seals, and other applications for which an oil-resistant elastomer is required. The elastomer, which has a density of 1.0 g/cm^3, can be formulated to have tensile

Table 6–2. Thermal, Physical, and Chemical Properties of Polysulfide Elastomer (Thiokol)(a)

Tensile strength, kPa............................7000
Elongation, %.....................................400
Hardness, Shore A 70
Specific gravity 1.34

Resistance to chemicals at 25 °C:(b)
    Nonoxidizing acids (20% H$_2$SO$_4$) ....S
    Oxidizing acids (10% HNO$_3$) ....Q
    Aqueous salt solutions (NaCl)..................S
    Aqueous alkalies (NaOH)......................Q
    Polar solvents (C$_2$H$_5$OH)........................S
    Nonpolar solvents (C$_6$H$_6$) ....S
    Water ....S

(a) Conversion tables appear in Appendix. (b) S, satisfactory; Q, questionable; U, unsatisfactory.

strengths in the range of 7000 to 20,000 kPa, an elongation from 400 to 600%, and a Shore durometer A hardness range of 40 to 95. Annually, 185,000 tons of NBR is produced worldwide, 75,000 tons of which is made in the United States. The formula for the repeating unit in NBR is shown below.

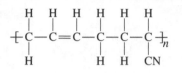

6.4 Polyacrylic Elastomers

Polyacrylic elastomers are copolymers of ethyl or butyl acrylate and comonomers such as 2-chlorovinyl ether, acrylonitrile, vinyl chloroacetate, or methacrylic acid. The first product of this type was developed by C. H. Fisher of USDA Eastern Regional Laboratories and was called Lastopar. Comparable products were developed by Goodrich chemists in the late 1940s. These are sold under the trade names Cyanacryl by American Cyanamid, Hycar 400 series by B. F. Goodrich, Thiacril by Thiokol, and Krynor by Polymer Corp.

Polyacrylic elastomers have excellent resistance to ozone, heat, and petroleum oils. Some of these polymers are cured (cross-linked) by long-chain amines or ammonium benzoate. The properties of polyacrylic elastomers depend on the composition. The tensile strength is in the range of 5000 to 125,000 kPa, the elongation is in the 100-to-400% range, and the Durometer A hardness is in the range of 40 to 90. These elastomers are used as oil hose, automotive gaskets, oil seals, and O-rings.

6.5 Polyfluorocarbon Elastomers

Many specialty solvent-resistant elastomers are produced by the copolymerization of fluorocarbon monomers. As we discussed in Section 5.10, polytetrafluoroethylene is intractable, but homopolymers containing fewer fluorine atoms on the repeating unit are more soluble and more flexible than PTFE. Also, as might be predicted from the relationship of properties to structure, copolymers of fluorocarbons are even more flexible and more readily processible than PTFE. The most widely used fluorocarbon elastomer is the copolymer vinylidene fluoride ($H_2C=CF_2$) and hexafluoropropylene (($F_2=CF$)(CF_3)). These elastomers ($VF_2 - HFP$), which are called Fluorel by 3M and Viton by duPont, are obtained by free-radical chain emulsion polymerization. The vinylidene fluoride (80)-hexafluoropropylene (20) has a density of 1.25 g/cm^3, a tensile strength of 17,000 kPa, an elongation of 300%, a Shore durometer A hardness of 68, and excellent resistance to heat. About 6000 tons of this elastomer is produced annually for use by the aerospace and industrial equipment industries. The structural formula for repeating units in VF_2—HFP is shown below.

$$-\!\!\!+\!CH_2-CF_2-CF_2-CF(CF_3)\!+\!\!\!-_n$$

Other temperature-resistant fluorocarbon elastomers include the copolymer of vinylidine fluoride (50) and chlorotrifluoroethylene (Kel F 5500); the copolymer of vinylidene fluoride, tetrafluoroethylene, and hexafluoropropylene (Viton G); and the copolymer of tetrafluoroethylene and perfluoromethyl vinyl ether (Kalrez). Fluoroacrylic elastomers such as 1,1-dihydroperfluorobutyl acrylate (FBA), which becomes brittle at -20 °C, fluorinated silicones, and hexafluoropentylene adipate are also available.

6.6 Polyurethanes (PUR)

As we stated in Section 5.7, snap-back fibers (spandex) can be produced by the reaction of a polyether or polyester, having terminal hydroxyl groups, with a diisocyanate. Although Bayer emphasized fibers in his polyurethane research in the 1930s, he did discover polyurethane (PUR) elastomers, which had poor tear resistance. However, high-performance elastomers have been produced and are still being produced by the reaction of diisocyanates with polyethers or polyesters having terminal hydroxyl groups.

These elastomers may be cast from a two-component system, and they may be cured by the addition of more diisocyanate, sulfur, or peroxy compounds. Pneumatic tire liners and long-wearing pneumatic tires are made by this technique. This same technique is used in reaction injection molding (RIM), but less flexible reactants are chosen for these reactions.

The most common glycols are butanediol adipate, polytetramethylene ether glycol, and polyoxypropylene glycol. The most widely used diisocyanates are TDI and MDI.

Cast polyurethane elastomers have high tensile strength (5000 kPa) and elongation (650%), excellent abrasion resistance, good oil resistance, and a Shore durometer A hardness of about 65. Moldable polyurethane elastomers, which are also available, have a density of 1.05 to 1.25 g/cm^3, a tensile strength of 14,000 to 100,000 kPa, an elongation of 100 to 650%, and a Shore durometer A hardness range of 65 to 80.

Millable elastomers (Vulcoprene A), which were the first PUR elastomers developed, are processed and fabricated by conventional techniques used in the rubber industry. A polyether-urethane based on 1,4-oxybutylene glycol and tolylene diisocyanate (TDI), Adiprene, can be cured by peroxy compounds or sulfur.

Thermoplastic elastomeric polyurethanes are based on adipic acid, 1,4-butanediol, and MDI (Estane), and on hydroxyl-terminated polyester diols and MDI (Texin). When heated, the latter are cured by a reaction with excess isocyanate on hydrogen atoms in PUR. This allophanate-type cross-linking reaction is illustrated by the following equation:

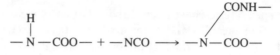

Allophanate linkage

Polyurethane elastomers are used for shoe soles, belts, wire and cable insulation, fuel hose, and O-rings, and as binders for solid propellants.

6.7 Silicones

Many pioneering organic chemists, including Berzelius, Friedel, and Kipping, experimented with organosilanes and siloxanes. Because he thought the latter were ketones, Kipping called them silicones. He also saw no practical use for these siloxanes.

The first silicone elastomer developed in the 1930s was polydimethylsiloxane. This polymer type is called DC-410 by Dow-Corning, SE-76 by General Electric, and W-95 by Union Carbide. These silicones may be cured (cross-linked) by peroxy compounds that serve as chain transfer agents and create polymerizable radicals. It is customary to use reinforcing fillers, such as colloidal silicas (Santocel, Cab-O-Sil, and Valcon).

Silicone elastomers have poor physical properties and are difficult to process. However, they are stable over a wide range of temperatures (-60 °C to 300 °C), are unaffected by ozone and hot oils, and have excellent electrical properties. Bouncing putty is produced by the reaction of dimethylsiloxane with iron (III) chloride and boron oxide (B_2O_3).

Silicones are used extensively as wire and cable insulation, gaskets, and in aerospace applications. Approximately 200,000 tons of silicone polymers are used annually worldwide. Silicone elastomers, called room-temperature vulcanizates (RTV), are used as potting compounds and for casting flexible articles. These siloxanes are cross-linked (cured) by the addition of organic metallic salts, such as dibutyltin laurate or tin (II) octoate. The structural formula for a repeating unit in a silicone elastomer is shown below.

Silicone elastomers may be used for long periods of time at 250 °C. Silicone elastomers with phenyl pendant groups are useful at relatively low temperatures (−70 °C).

6.8 Polyphosphazenes

Phosphonitrilic compounds were investigated by Liebig in 1834, and Stokes in 1835 prepared phosphonitrilic chloride by heating a mixture of phosphorus pentachloride (PCl_5) and ammonium chloride (NH_4Cl) in an inert solvent.

Schenck and Rauer produced polyphosphonitrilic chloride elastomers in 1924 by heating phosphonitrilic chloride ($N=PCl_2$) at 300 °C. Since this polymer, like its precursor (PCl_5), is hydrolyzed in moist air, the chlorine atoms were replaced by alkoxides, aryloxides, or amines. The resultant polymers, called phosphazenes, are resistant to hydrolysis, have a low glass transition temperature (−66 °C), and retain their flexibility up to their melting point (240 °C). The formula for the repeating unit of poly (bis(trifluoroethoxy)) phosphazene is shown below.

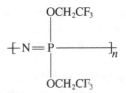

A clear, water-soluble thermoplastic is obtained when polyphosphonitrile chloride is reacted with methylamine (CH_3NH_2). The introduction of more than one type of substituent on the phosphorus atom reduces the crystallinity and enhances the elasticity of phosphazenes. Phosphazenes with residual chloride pendant groups can be cross-linked by the addition of moisture.

Although phosphazene elastomers are useful over a wide range of temperatures, they depolymerize at temperatures above 200 °C. Polyphosphazenes have been used as fuel lines, hoses, gaskets, O-rings, and foams.

6.9 Polyether Elastomers

Elastomers under the trade name Hydrin are produced by Hercules by the polymerization of epichlorohydrin and propylene oxide, respectively. The Hydrin elastomer is cured by the addition of substances that react with chlorine, and another polyether marketed by Hercules under the trade name Parel is cured by sulfur-containing accelerators. These elastomers have specific gravities of 1.36 and 1.01, respectively. Because of their excellent oil resistance, they are used for automotive motor moldings.

86 Polymers for Engineering Applications

6.10 Thermoplastic Elastomers

As stated in Section 4.1.1, Hevea rubber, which cold flows excessively, is an elastomer, but has limited utility until it is cross-linked (vulcanized). In contrast, many thermoplastics are nontacky and do not cold flow excessively, but they are not elastic. Those who understand polymer structure would recognize that these diverse properties could be combined in one macromolecule, but such knowledge was not available until relatively recently. In 1838, however, Charles Mackintosh did incorporate elastic and plastic properties into a single composite by making a sandwich (laminate) of cloth and unvulcanized rubber.

More recently, chemists have incorporated elastic and plastic properties into a macromolecule by synthesizing block or graft copolymers. When a series of rigid repeating units (domains) is joined to a series of elastic repeating units in a polymer chain, the former act like cross-links and prevent cold flow of the elastomeric domains.

6.10.1 Styrene-Butadiene Block Copolymers

AB and ABA block copolymers of this type, called plastomers, have been made from styrene and butadiene. The styrene(A)-butadiene(B) block copolymers are produced by Phillips Petroleum under the trade name Solprene. ABA-styrene-butadiene-styrene block copolymers are produced by Shell Oil under the trade name Kraton. The tensile strength and hardness of these copolymers is a function of the styrene content. Tensile strength is independent of molecular weight at room temperature, but the tensile strength increases as the molecular weight is increased at higher temperatures, such as at 60 °C.

An AB copolymer with a 40-to-60 ratio of styrene-butadiene block copolymer will have a density of 0.55 g/cm^3, a tensile strength of 28,000 kPa, an elongation of 700%, and a Shore A hardness of 93. SBS block copolymers are characterized by glass transition temperatures (T_g) of 100 °C and −60 °C for the incompatible S and B blocks, respectively.

ABA block copolymers may be hydrogenated to produce saturated polymers. The saturated SBS copolymers with 40% styrene have a tensile strength of 28,000 kPa, an elongation of 750%, and a Shore A hardness of 92 at 25 °C. AB or ABA block copolymers of styrene and butadiene may be used in place of vulcanized NR in applications such as automotive parts, hose, wire and cable coating, footwear, and gaskets. The properties of a typical SBS block copolymer are listed in Table 6–3, and a repeating unit in these blocks is shown below.

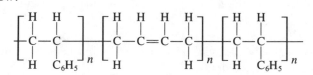

Table 6–3. Thermal, Physical, and Chemical Properties of a Typical Styrene-Butadiene-Styrene Block Terpolymer(a)

Tensile strength, kPa...........................39,000
Elongation, %......................................600
Hardness, Shore A 60
Specific gravity 0.9
Dielectric constant...............................3.0

Resistance to chemicals at 25 °C:(b)
Nonoxidizing acids (20% H_2SO_4) .... S
Oxidizing acids (10% HNO_3) ....Q
Aqueous salt solutions (NaCl)..................S
Aqueous alkalies (NaOH)........................S
Polar solvents (C_2H_5OH).........................S
Nonpolar solvents (C_6H_6) ....U
Water .... S

(a) Conversion tables appear in Appendix. (b) S, satisfactory; Q, questionable; U, unsatisfactory.

6.10.2 Polyester Thermoplastic Elastomers

Another type of block copolymer that fills the gap between elastomers and rigid plastics is segmental copolyester rubber, which is marketed by duPont under the trade name Hytrel. These thermoplastic elastomers (TPE) are produced by the transesterification of dimethyl terephthalate polytetramethylene ether glycol (PTMEG) and 1,4-butane diol (4GT). The resulting products are random block copolymers consisting of crystalline 4GT hard segments and soft segments of PTMEG.

The crystalline phase of these block copolymers acts as a thermally reversible cross-link and the amorphous phase contributes to elasticity. The modulus and hardness are functions of the 4GT content. These block copolymers are used as hose, wire coating, seals, and gaskets. The properties of these thermoplastic elastomers are shown in Table 6–4.

6.10.3 Thermoplastic Polyurethane Elastomers (TPU)

Thermoplastic polyurethane elastomers (TPU) consist of soft (low T_g) blocks of moderate-molecular-weight aliphatic polyethers or polyesters and hard blocks formed by the reaction of diisocyanates with low-molecular-weight diols.

These polymers are molded or extruded to produce automotive parts, hose, gears, shoe soles, and tire chains.

The properties of these TPUs are summarized in Table 6–5.

6.10.4 Silicone Block Copolymers

Copolymers with relatively short blocks of polystyrene and polydimethylsiloxane are also available commercially from Dow-Corning. Unlike the AB and

Table 6–4. Thermal, Physical, and Chemical Properties of a Typical Polyester Thermoplastic Elastomer(a)

Tensile strength, kPa......................... 39,000
Elongation, %..................................... 350
Hardness, Shore D 72
Specific gravity 1.25
Izod notched impact strength,
  J/cm 2.1
Heat deflection temperature at
  500 kPa, °C...................................... 166
Softening point, Vicat °C 203
Coefficient of linear expan-
  sion, cm/cm/°C $\times$ 10^{-5} 21
Water absorption, % 0.3

Resistance to chemicals at 25 °C:(b)
  Nonoxidizing acids (20% H_2SO_4) .... S
  Oxidizing acids (10% HNO_3) .... U
  Aqueous salt solutions (NaCl)................. S
  Aqueous alkalies (NaOH)...................... Q
  Polar solvents (C_2H_5OH)....................... S
  Nonpolar solvents (C_6H_6) .... Q

(a) Conversion tables appear in Appendix. (b) S, satisfactory; Q, questionable; U, unsatisfactory.

Table 6–5. Thermal, Physical, and Chemical Properties of a Typical Thermoplastic Polyurethane Elastomer (TPU)(a)

Heat deflection temperature
  at 1820 kPa, °C..................................70
Coefficient of linear expansion,
  cm/cm/°C $\times$ 10^{-5} 15
Tensile strength, kPa....................... 20,000
Elongation, %..................................... 600
Specific gravity 1.25
Hardness, Shore A 80

Resistance to chemicals at 25 °C:(b)
  Nonoxidizing acids (20% H_2SO_4) .... Q
  Oxidizing acids (10% HNO_3) .... U
  Aqueous salt solutions (NaCl)................. S
  Aqueous alkalies (NaOH)...................... Q
  Polar solvents (C_2H_5OH)....................... U
  Nonpolar solvents (C_6H_6) .... Q
  Water .... S

(a) Conversion tables appear in Appendix. (b) S, satisfactory; Q, questionable; U, unsatisfactory.

ABA block copolymers, these products have as many as six or eight alternate blocks. The polystyrene blocks have $\overline{\mathrm{DP}}$s of at least 80 and these domains control the moduli and yield points of these polymers. A block with the $\overline{\mathrm{DP}}$ of 1200 of polystyrene will have a tensile strength of 1000 kPa and an elongation of 350%.

Clear block copolymers of polydimethylsiloxane and polycarbonate (40%), which are also available from General Electric, have a tensile strength of 16,000 kPa. Blocks of polydimethylsiloxane (50%) and polysulfone, which are available from Union Carbide, have properties similar to those of TPUs. For example, a copolymer consisting of 55% polydimethylsiloxane would have a tensile strength of 18,000 kPa and an elongation of 350%. Thus, like other block copolymers, these products will have separate T_g's for each domain that, in turn, are 120 °C and 160 °C. A section of a typical silicone-polysulfone block copolymer is shown below.

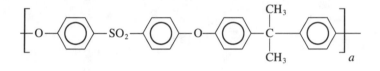

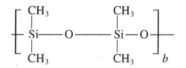

High-Performance Coatings

As we stated in Section 4.3, most coatings and substrates are high-performance systems. Thus, when properly applied on a substrate, even unsophisticated coatings such as oleoresinous paints, cellulose nitrate lacquers, and some alkyd and acrylic coatings yield high-performance systems. Because of good resistance to permeation by the corrosive environment, these and more sophisticated coatings protect surfaces that would be chemically attacked if unprotected. It is also important to note that plastic surfaces are often coated with metals to ensure electrical conductivity or to provide a metallic appearance.

7.1 Rubber and Its Derivatives

Provided that a metal surface is primed with a polymer such as cyclized natural rubber (isomerized rubber), compounded natural rubber (*Hevea*) latex may be applied to protect the metal surface. Built-up latex rubber coatings have been used to protect the metal tanks used for hot acid pickling of steel. These linings are protected by acid-proof brick when the temperature exceeds 50 °C. The properties of the film in the rubber coatings are shown in Table 7–1.

7.1.1 Chlorinated Rubber

Chlorinated rubber with a chlorine content of 67% and a density of 1.6 g/cm³ has been available for over 50 years. It continues to be marketed under the trade names Parlon (Hercules), Alloprene (ICI), and Tornesit and Pergut (Bayer). This polymer is insoluble in aliphatic hydrocarbons, such as heptane,

Table 7–1. Thermal, Physical, and Chemical Properties of Carbon-Filled, Vulcanized (*Hevea*) Rubber Coatings (Properties of Deposited Film)(a)

Maximum resistance to continuous
  heat, °C... 50
Coefficient of linear expansion,
  cm/cm/°C × 10^{-5} 15
Tensile strength, kPa..........................30,000
Elongation, % 50
Hardness, Shore D 80
Dielectric constant...................................3.2
Water absorption, % 2.0

Resistance to chemicals at 25 °C:(b)
  Nonoxidizing acids (20% H_2SO_4) .... S
  Oxidizing acids (10% HNO_3) ....U
  Aqueous salt solutions (NaCl)................... S
  Aqueous alkalies (NaOH)......................... S
  Polar solvents (C_2H_5OH)......................... S
  Nonpolar solvents (C_6H_6) ....U
  Water .... S

(a) Conversion tables appear in Appendix. (b) S, satisfactory; Q, questionable; U, unsatisfactory.

and alkanols, such as ethyl alcohol, but is soluble in most other organic solvents. It adheres well to portland cement and metal surfaces and is resistant to most nonoxidizing acids and alkalies.

7.1.2 Cyclized Rubber

Cyclized or isomerized rubber (Pliolite) is produced by the chlorostannic acid (H_2SnCl_6) isomerization of natural rubber. This polymer is compatible with paraffin wax and many hydrocarbon resins. It is soluble in both aliphatic and aromatic hydrocarbon solvents and is usually applied as a solution in toluene.

7.2 Other Hydrocarbon Resins

Bitumens have been used for centuries as protective coatings. Yet in spite of their utility, they are not classified as high-performance coatings. Several more sophisticated hydrocarbon resins providing quality protective coatings are obtained by the distillation of coal tar and petroleum crudes. Coumarone-indene resins (Cumar) are obtained by the low-temperature cationic polymerization of these unsaturated hydrocarbon monomers. These polymer coatings are usually deposited from toluene solutions. They are soluble in many organic solvents and are useful at temperatures up to 50 °C. These polymers adhere well to metal surfaces and are resistant to nonoxidizing acids and alkalies.

Comparable resins from petroleum (Piccopale) and turpentine (Piccolyte) are also produced by cationic polymerization of the unsaturated monomers. These polymers are also used as coatings and as modifiers for other coating systems. The formulas for the repeating units for polyindene and polycoumarone are shown below.

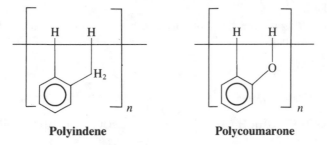

Polyindene   Polycoumarone

7.3 Alkyds and Polyester

Polyester coatings, obtained by the esterification of glycerol ($CH_2C(OH)CH \cdot (OH)CH_2(OH)$) and phthalic anhydride ($C_6H_4C_2O_3$), were produced by W. Smith in the early 1900s, and marketed by General Electric under the trade name Glyptal. They were used as insulating varnishes. Because the secondary hydroxyl group in glycerol is less reactive than the two primary hydroxyl groups, glycerol is bifunctional at moderate esterification temperatures and yields a linear prepolymer that has a free hydroxyl group on each repeating unit. The hydroxyl reacts at elevated temperatures when the solvent-free coating is baked after application. (Some information on polyesters was also provided in Section 4.4.3.)

In 1927, Kienle substituted the difunctional ethylene glycol for glycerol and obtained additional functionality by the incorporation of unsaturated oils. The unsaturated prepolymers obtained, which he called alkyds, can be cross-linked by oxygen in the presence of heavy metal catalysts (driers), much like the conventional oleoresinous paints. Kienle used the terms "short," "medium," and "long" oil alkyds to show the relative amounts of unsaturated oils used in the unsaturated prepolymer.

Subsequently, Kienle used vegetable oils (drying oils) in place of unsaturated fatty acids by use of alcoholysis techniques. The alkyds were also modified by the addition of phenolic resins; the phthalic anhydride was replaced, at least in part, by aliphatic dicarboxylic acid; and the ethylene glycol and glycerol were replaced by pentaerythritol or sorbitol, which have four and six hydroxyl groups in the molecules.

In 1940, Ellis and Rust substituted maleic anhydride in place of part of the phthalic anhydride and produced a coating by dissolving the unsaturated polyester, produced by condensation polymerization, in styrene. The prepolymer solution was cured after application by free-radical chain polymerization. This unsaturated polyester-styrene solution was used for coating metals, wood, and concrete, and for the impregnation of fibrous glass mats to produce reinforced plastics (FRP).

The formulas for the reactants used in producing alkyds and other polyesters are shown below.

$HO-CH_2-CH_2-OH$ $(HOCH_2)_4C$

Ethylene glycol   **Pentaerythritol**

$HOCH_2CH(OH)CH_2OH$ $HOCH_2(CHOH)_4CH_2OH$

Glycerol   **Sorbitol**

Glycols

$HOOC(CH_2)_4COOH$

Adipic acid   **Phthalic anhydride**

$HOOC(CH_2)_8COOH$ $H_3C(CH_2)_7CH{=}CH(CH_2)_7COOH$

Sebacic acid   **Oleic acid**

                     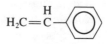

Maleic anhydride   **Styrene**

Fumaric acid   **Maleic acid**

The properties of a typical alkyd coating are shown in Table 7–2.

Chlorendic anhydride (HET) and tetrachlorophthalic anhydride are highly chlorinated anhydrides used to produce flame-retardant polyester coatings. Another polyester, called "vinyl ester" (Derakane), is an acrylic ester of epoxidized *bis*-phenol-A. Fumaric acid esters of epoxidized *bis*-phenol-A are also used for the production of high-performance unsaturated polyesters. Fumaric acid is a geometrical trans acid of maleic acid, which is a cis acid. The formula for the repeating unit in vinyl esters (Derakane) is shown below.

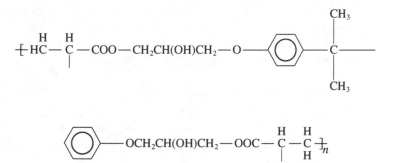

Table 7–2. Thermal, Physical, and Chemical Properties of a Typical Alkyd Coating(a)

Maximum resistance to continuous
  heat, °C... 90
Coefficient of linear expansion,
  cm/cm/°C × 10^{-5} 5
Tensile strength, kPa..........................35,000
Elongation, % 65
Hardness, Shore D 80
Specific gravity 1.2
Dielectric constant...................................4
Water absorption, % 2

Resistance to chemicals at 25 °C:(b)
  Nonoxidizing acids (20% H_2SO_4) .... S
  Oxidizing acids (10% H_2O_3) ....U
  Aqueous salt solutions (NaCl).................. S
  Aqueous alkalies (NaOH)........................Q
  Polar solvents (C_2H_5OH)........................ S
  Nonpolar solvents (C_6H_6) ....Q
  Water .... S

(a) Conversion tables appear in Appendix. (b) S, satisfactory; Q, questionable; U, unsatisfactory.

7.4 Phenolic and Amine Resins

Some information on phenolic, urea-formaldehyde, and melamine-formaldehyde resins was provided in Section 4.4.

7.4.1 Phenol Formaldehyde Coatings

The resol resins produced by the condensation of phenol and formaldehyde under alkaline conditions have been used as baked high-performance coatings. Linear oil-soluble phenolic coatings, which do not require baking, are reaction products of formaldehyde and substituted phenols, such as p-octylphenol and p-phenylphenol. Unsaturated vegetable oils may be included in the formulation of these albertol-type coatings. High-performance adhesives are produced when these soluble phenolic resins are admixed with neoprene.

7.4.2 Urea-Formaldehyde Coatings

Urea-formaldehyde resins (UF) under the trade name Beetle were used as molding resins in England in 1926. Urea-formaldehyde adhesives under the trade name Kaurit were commercialized in Germany in 1933. The utility of these UF coating resins was improved when the reactants were condensed, under acid conditions, in the presence of butanol.

UF resins have been used as adhesives for chipboard, paper in paper-based laminates, and in furniture; as coatings, for the production of wet-strength paper, and wash-and-wear textiles. Butanol-etherified UF resins are used as modifiers for alkyd and cellulose nitrate resins in furniture coatings.

7.4.3 Melamine-Formaldehyde Coatings

Melamine-formaldehyde resins (MF), like UF resins, are colorless and are used in place of phenolic resins, where their dark characteristic color and lack of alkali resistance are not acceptable. MF coatings have better resistance to moisture, elevated temperatures, and corrosives than UF coatings for similar applications.

7.5 Acrylic Coatings

Acrylic coatings, based on blends of polymethyl methacrylate and polyethyl acrylate, are transparent and durable. They are usually applied over epoxy-resin-based primers. The thermoplastic acrylic coatings are used as protective top coats for PVC films and plastics, for protection of aluminum surfaces, and as surface coatings for plastics before they are metallized. The resistance of acrylic coatings to water spotting and etching may be overcome by using methacrylic esters of higher alcohols, such as octyl methacrylate. The chemical resistance of the methacrylates is superior to that of the acrylates.

Thermosetting acrylic coatings are produced by copolymerization of the monofunctional acrylic monomers with acrylamide, hydroxyalkyl acrylates, glycidyl acrylates, or methacrylates. These thermoset acrylic coatings are used as finishes for appliances, copper, aluminum siding, and automobile bodies. When photoinitiators, such as benzoin ethers, acetophenone derivatives, acyloxime esters, and benzols, are present, these coatings may be cured by ultraviolet radiation.

7.6 Epoxy Resins (EP)

Epoxy resins (EP), under the name of ethyoxylene resins, were patented in Europe by P. Schlack in 1939. These resins, which are produced by the condensation of *bis*-phenol-A (BPA) and epichlorohydrin (ECH), contain terminal epoxy groups and multiple hydroxyl groups on the polymer chain.

The linear polymer may be cross-linked (cured) at ordinary temperatures by the addition of polyfunctional amines, such as diethylenetriamine (DTA), which reacts with the epoxy end groups. Cross-links may also be introduced by heating EP with cyclic anhydrides, such as phthalic anhydride, which react with the hydroxyl groups present on each repeating unit in the polymer chain. Because the amine curing agents may cause dermatitis, polyamine derivatives produced by the reaction of polyfunctional amines and dimerized vegetable oils are also used as room-temperature curing agents.

Because of the presence of polar groups, EP resins are excellent adhesives for metals, elastomers, wood, and glass. EP resins are widely used as coatings and as components of other coatings, such as acrylic, phenolic, and urea resins and coal tar. They are also used in place of polyester resins for fiber-reinforced plastics.

EP resins are marketed as two-package resin systems under the trade names Araldite, Epikote, Epon, and Epoxylite. Over 150,000 tons of epoxy resins are used annually in the United States, and their use is growing rapidly.

The properties of a typical epoxy resin coating are shown in Table 7–3, and the formula for an uncured epoxy resin is shown on the following page.

Epoxy-like resins, which lack the terminal epoxide groups, are called phenoxy resins. These linear polymers are used as surface coatings and adhesives and can be cured by cross-linking of the hydroxyl groups, in the repeating units, by polyisocyanates, amino resins, or phenolic resins.

Table 7–3. Thermal, Physical, and Chemical Properties of a Typical Epoxy Resin Coating(a)

Heat deflection temperature at
  1820 kPa, °C 140
Maximum resistance to continuous
  heat, °C 130
Coefficient of linear expansion,
  cm/cm/°C $\times$ 10^{-5} 5
Tensile strength, kPa 50,000
Elongation, % 5
Hardness, Shore D .... .85
Specific gravity 112
Water absorption, % 1
Dielectric constant 4

Resistance to chemicals at 25 °C:(b)
  Nonoxidizing acids (20% H_2SO_4) .... S
  Oxidizing acids (10% HNO_3) .... U
  Aqueous salt solutions (NaCl) .... S
  Aqueous alkalies (NaOH) .... S
  Polar solvents (C_2H_5OH) .... S
  Nonpolar solvents (C_6H_6) .... S
  Water .... S

(a) Conversion tables appear in Appendix. (b) S, satisfactory; Q, questionable; U, unsatisfactory.

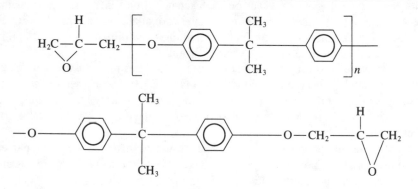

7.7 Polyurethane Coatings (PUR)

Some information on polyurethane (PUR) was provided in Sections 5.7 and 6.6. These resins, which were introduced by O. Bayer in 1937, are produced by the reaction of a diisocyanate and a diol. The coatings may be urethane oils, moisture-cured PUR, two-pot systems, and blocked PUR.

The urethane oils, in which the diisocyanate is reacted with partially hydrolyzed unsaturated drying oils, which cure by heavy metal-catalyzed cross-linking in the presence of air, were the first commercial PUR coatings. The isocyanate reacts with the hydroxyl groups on the mono- and diglycerides produced by the hydrolysis reaction, as shown below. These polymers provide tough, clear coatings.

$$
\begin{array}{ccc}
CH_2O\!-\!UA & & CH_2OUA \qquad\qquad CH_2OUA \\
| & & | \qquad\qquad\qquad | \\
2\,CHOH \quad + OCN\!-\!R\!-\!NCO \longrightarrow & HC\!-\!OOCNHRNHCOO\!-\!CH \\
| & & | \qquad\qquad\qquad | \\
CH_2O\!-\!UA & & CH_2OUA \qquad\qquad CH_2OUA
\end{array}
$$

**Diglyceride of
unsaturated acids (UA)**   **Urethane**

The formula for diphenylmethane-4,4-diisocyanate (MDI) is shown below.

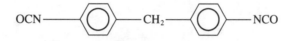

$$OCN\text{———}\bigcirc\text{———}CH_2\text{———}\bigcirc\text{———}NCO$$

Polyurethane, produced from an excess of diisocyanate, such as diphenyl-methane-4,4-diisocyanate (MDI), will react, in the presence of tertiary amine catalysts, with moisture from the atmosphere. The carbon dioxide released in this reaction will pass through thin films but will produce bubbles in thick films. The equation for the reaction between a diisocyanate and water is shown below.

$$
\overset{\textstyle H}{\underset{\textstyle |}{}}
$$
$$OCN\!-\!R\!-\!NCO + H_2O \longrightarrow OCN\!-\!RNCOOH \longrightarrow CO_2 + OCNRNH_2$$

The amine group may react further with isocyanates to produce a urea, as shown by the following equation:

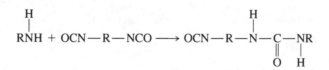

$$
\overset{H}{\underset{|}{RNH}} + OCN\!-\!R\!-\!NCO \longrightarrow OCN\!-\!R\!-\!\overset{H}{\underset{|}{N}}\!-\!\overset{}{\underset{\substack{\| \\ O}}{C}}\!-\!\overset{}{\underset{H}{N}}R
$$

Urea derivative

Moisture-cured, thick PUR coatings are obtained by use of multiple coats applied as separate layers after the evolution of carbon dioxide is complete in the previous layer.

Polyurethane coatings may be produced *in situ* by admixing a polyether or polyester urethane having terminal hydroxyl groups with a PUR prepolymer having isocyanate terminal groups. When more than two hydroxyl groups are present in the reactant molecules, cross-linking will occur. Since no by-products are produced, thick coatings may be deposited from these two-package systems.

Baked coatings can be produced from blocked diisocyanates, such as phenol adducts, which when heated release the volatile phenol and the diisocyanate. The latter then reacts with diols in the formulation. Ethyl malonate, phenolic resins, caprolactam, and ketoximes have been used to produce blocked diisocyanates. Commercial blocked isocyanates are marketed by duPont (Hylene), Mobay (Mondurs), and Dow (Isonate 123P). Aminimide blocked diisocyanates are available from Ashland. In addition to their use as high-performance coatings, polyurethanes are also excellent adhesives.

The properties of PUR coatings from a two-package system are shown in Table 7–4.

7.8 Silicone Coatings

The room-temperature vulcanizable (RTV) potting compositions were discussed in Section 6.7. These and other siloxanes (silicones) used as coatings are characterized by excellent resistance to thermal and oxidative degradation, high water repellency, retention of good physical properties over a wide range of temperatures, flexibility because of the so-called free rotation of the Si—O bulky groups in the polymer chain, low toxicity, and incompatibility with most other coatings.

The polymerization of silanes involves the hydrolysis of chloroalkylsilanes or methoxyalkylsilanes, followed by neutralization and curing, as shown in the following equation:

$$
n\;Cl-\underset{\underset{CH_3}{|}}{\overset{\overset{CH_3}{|}}{Si}}-Cl \quad \underline{+H_2O} \quad n[HO-\underset{\underset{CH_3}{|}}{\overset{\overset{CH_3}{|}}{Si}}-OH] \quad \underline{-H_2O} \quad \left[\!\!-\underset{\underset{CH_3}{|}}{\overset{\overset{CH_3}{|}}{Si}}-O-\!\!\right]_n
$$

Dichlorodimethyl silane   **Silicone**

A small amount of trichloromethylsilane is usually present to cure (cross-link) the linear polymer produced by the hydrolysis of a difunctional dichlorodimethylsilane.

Table 7–4. Thermal, Physical, and Chemical Properties of a Typical PUR Coating Deposited From a Two-Package System(a)

Heat deflection temperature at
  1820 kPa, °C 75
Maximum resistance to continuous
  heat, °C... 70
Coefficient of linear expansion,
  cm/cm/°C × 10^{-5} 15.0
Tensile strength, kPa........................... 6,890
Elongation, % 200
Hardness, Shore D50
Specific gravity 1.2
Dielectric constant.................................... 6

Resistance to chemicals at 25 °C:(b)
  Nonoxidizing acids (20% H_2SO_4) .... Q
  Oxidizing acids (10% HNO_3) .... U
  Aqueous salt solutions (NaCl)................... S
  Aqueous alkaline solutions (NaCH)............ Q
  Polar solvents (C_2H_5OH)......................... U
  Nonpolar solvents (C_6H_6) .... Q
  Water .... S

(a) Conversion tables appear in Appendix. (b) S, satisfactory; Q, questionable; U, unsatisfactory.

Because of the high polarity of siloxanes and the low intermolecular forces between polymer chains, polysiloxanes are permeable to many gases and liquids. Because of their high cost, they are often modified by the addition of other polymers, such as alkyds. The principal use for silicone coatings is in heat- and water-resistant coatings, electrical insulation, and water-repellent masonry coatings. Since silicones are transparent to ultraviolet radiation, they are more resistant to weathering than many other coatings.

The properties of typical silicone coatings are presented in Table 7–5.

7.9 Poly-*p*-Xylylene

Poly-*p*-xylylene (Parylene) is a specialty high-performance coating resin that must be applied in a special vacuum chamber in which the dimer (di-*p*-xylene) is pyrolyzed at 600 °C in a vacuum to produce *p*-xylylene, which polymerizes spontaneously when cooled. The precursor to the dimer (di-*p*-xylene) is a monomer that, when heated with steam at 900 °C, forms the dimer. Poly-monochloro- and polydichloro-*p*-xylylenes are also available. The melting points of these resins are in the 290-to-400 °C range.

The weather resistance of these coatings is poor. They are not resistant to ultraviolet radiation, but they are serviceable over a wide range of tempera-

Table 7–5. Thermal, Physical, and Chemical Properties of Typical Silicone Coatings(a)

Maximum resistance to continuous
  heat, °C..250
Coefficient of linear expansion,
  cm/cm/°C × 10^{-5} 40
Tensile strength, kPa..........................5,000
Elongation, % 200
Hardness, Shore A 65
Specific gravity 1.2
Dielectric constant..................................3
Water absorption, % 0.2

Resistance to chemicals at 25 °C:(b)
  Nonoxidizing acids (20% H_2SO_4) ....Q
  Oxidizing acids (10% HNO_3) ....U
  Aqueous salt solutions (NaCl)...................S
  Aqueous alkalies (NaCl) ....S
  Polar solvents (C_2H_5OH).........................S
  Nonpolar solvents (C_6H_6) ....Q
  Water ....S

(a) Conversion tables appear in Appendix. (b) S, satisfactory; Q, questionable; U, unsatisfactory.

tures in inert atmospheres. Poly-*p*-xylylenes are used extensively in the protection of hybrid circuits. The equation for the formation of poly-*p*-xylylene is shown below.

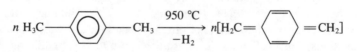

p-Xylene

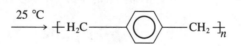

Poly-*p*-xylylene

Functional Polymers and Specialty Polymers

Most general-purpose and high-performance polymers are molded and extruded, and some consumers may mistakenly believe that the end uses of polymers are limited to molded parts and extrudates. Fortunately, polymers are more versatile and perform many other functions, which we address in this chapter.

8.1 Conductive Polymers

Organic polymers, like other organic compounds, are usually made up of carbon atoms joined together by covalent bonds, which by definition are nonconductive. Hence, many moisture-resistant polymers are used as electrical and thermal insulators. Actually, polyolefins and polyfluorocarbons are among the best insulators available.

Nevertheless, it has been recognized that graphite, which consists of sheets of carbon atoms with delocalized electrons, is electrically conductive. Consequently, graphite fibers are used as conductors in electric blankets and other low-grade heating elements.

Since electrical conductivity is essential to protect against electromagnetic interference (EMI), conductive polymers have been produced by the addition of stainless steel wire or other metals in the form of flakes or powders.

The first synthetic nonmetal-filled conductor was polysulfur nitride $(SN)_n$, which was synthesized by Burt in 1910 and becomes a superconductor at 0.3 K. Polyphthalocyanine, polythiophene, polypyrrole, polyphenylene, polyphenylene sulfide, and polyacetylene have a small degree of conductivity, and this effect is enhanced by the addition of dopants, such as arsenic pentafluoride. The conductivity of polyacetylene can be increased by 13 orders of magnitude by doping. Accordingly, these polymers and other conductive doped polymers have been called "organic metals."

Shirakawa and Ikeda in 1971 prepared free-standing film from polyacetylene with a metallic luster. *cis*-Polyacetylene was obtained by the low-temperature polymerization of acetylene in the presence of Ziegler-type catalysts. Blue-colored *trans*-polyacetylene was obtained by the polymerization of acetylene at high temperatures. The conductivities are 10^{-10} and 10^{-6} (ohm-cm)$^{-1}$ for the cis and trans isomers, respectively. The cis polymer is nonphotoconductive, but the trans polymer is photoconductive. The formulas for the repeating units in polysulfur nitride and polyacetylene are shown below.

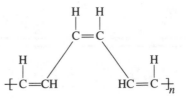

Polysulfur nitride   *cis*-Polyacetylene

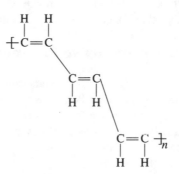

trans-Polyacetylene

As a result of the formation of conjugated double bonds, the conductivity of PVC and PVA increases when they are heated at high temperatures. Batteries have been produced by immersing polyacetylene film in a propylene carbonate solution of lithium perchlorate ($LiClO_4$). The composition of the film in the charged battery is $+CH(ClO_4)_{0.06}+_n$.

Polymers with conductivity greater than 10^{-8} (ohm-cm)$^{-1}$ do not store electrostatic charges when rubbed against a dissimilar surface. However, when rubber and nonconducting polymers, such as polystyrene, accumulate electrostatic charges, they retain them over long periods of time. The ancient Greeks developed a triboelectric series based on rubbing glass, amber, and other non-conductors. Antistats, which attract moisture that dissipates the electrostatic charge in polymers, are added to polymer films, for example, to prevent the accumulation of dust and other undesirable particles.

Nonconducting polymers may also be charged to produce "electrets" by the application of an electric field at a temperature of about 35 K above T_g and allowing the polymer to cool while still in the electric field. These charges are retained for several months.

8.2 Photoconductive Polymers

Photoconductors, which generate current in the presence of light, consist of very thin layers of vitreous selenium and have been the accepted materials for efficient transport of electrons, but polymeric photoconductors are also being used. The requirements for polymeric photoconduction are low, dark conductivity and good mobility of carriers generated by the absorption of light.

Polymers such as poly-N-vinylcarbazole (PVCA), PVCA with trinitrofluorenone, or molecularly doped PVCA have a strong intrinsic optical absorption in the ultraviolet range but not in the visible range. This deficiency can be overcome by the addition of an absorbing dye, such as trinitrofluorenone or a thin film of selenium. The addition of trinitrofluorenone to polycarbonates also produces photoconductivity. Polyvinyl cinnamate, produced by the esterification of polyvinyl alcohol, cross-links when exposed to ultraviolet radiation. The formula for the repeating unit in polyvinylcarbazole is shown below.

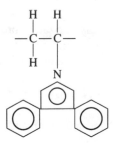

PVCA containing a Lewis acid and an optical sensitizer (dye) is used in xerography.

8.3 Piezoelectrical and Pyroelectrical Polymers

The strong tendency to generate current under pressure or heat is called piezo- or pyroelectricity. Electrets, such as polyvinylidene fluoride film (PVDF), were discovered in the early 1970s. Properly prepared PVDF film is used as an acoustic device in high-frequency speakers and headphones, ultrasonic transducers, and acoustic emission devices. Pyroelectric polymer films have been used as optical detectors for infrared radiation and millimeter waves. The piezoelectricity of nylon-11 is about 50% that of PVDF.

8.4 Ion-Exchange Resins

The reversible interchange of ions between a solid and a liquid is essential for the deionization of water and products in aqueous solutions. Aluminum silicates called zeolites, which have been used for ion exchange, are being replaced by ion-exchange resins. Cation exchange occurs when the functional groups on the resin are negatively charged, and anion exchange occurs when the groups are positively charged. It is customary to place the ion exchanger in a column and to pass the solution through the packed column.

The ion-exchange resins are usually cross-linked polymers with positively or negatively charged functional groups. The pioneer ion-exchange resin was phenol formaldehyde (PF), in which the phenolic hydroxyl group absorbed cations. The ion-exchange function of these resins was improved by placing sulfonic acid groups on the resins and by the condensation of phenol and formaldehyde in the presence of sodium sulfite. Anion-exchange PF resins were produced by the addition of amines to the phenol and formaldehyde before condensation.

Cation-exchange resins are also produced by the reaction of chlorosulfonic acid with polystyrene beads cross-linked (with divinylbenzene, or DVB). The reaction of chloromethylated cross-linked polystyrene with a tertiary amine produces an anion-exchange resin (DOWEX). The size of the beads may be controlled in the suspension polymerization process, and the size of the pores in the beads may be controlled by the extent of cross-linking. Permaselective membranes are made from ion-exchange resins.

8.5 Microencapsulated and Controlled-Release Polymers

Carbon copies of typed or handwritten manuscripts are no longer made by placing carbon paper between two sheets of paper. Carbonless-copy business forms contain a microencapsulated crystal violet lactone deposited in a thin film on the underside of the top sheet and an undersheet coated topside with microencapsulated clay. The two reactants are released when the microcapsules are broken by the pressure of a pencil or ballpoint pen.

Anaerobic adhesives can be coated on bolts and released when the bolts are screwed into place. Polytetraethylene glycol dimethacrylate, which polymerizes with a redox initiator system in the absence of oxygen, is an example of an anaerobic adhesive. Water may also be microencapsulated and permitted to react with gypsum when rods are rammed into place. The formula for the repeating unit in polytetraethylene glycol dimethacrylate is shown below.

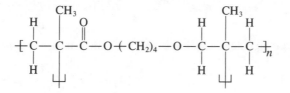

Many controlled-release pharmaceuticals, including aspirin and antibiotics, depend on microencapsulation for their release. Membranes that become permeable in the presence of moisture from the skin are used as patches containing drugs, such as glyceryl nitrate for angina pectoris therapy.

Other controlled-release techniques using polymers include ocular therapeutic systems, which provide an alternative to eyedrops or ointments for eye medications, including the slow release of pilocarpine for lowering ocular pressure in glaucoma; transdermal patches, which release scopolamine for the control of motion sickness; transdermal patches for the controlled release of estradiol for the relief of menopausal symptoms; transdermal patches for the release of phenyl propanolamine for the suppression of appetite; and controlled release systems for the release of biocides in agriculture. Many additional controlled-release systems are being studied by the U.S. Food and Drug Administration.

8.6 Oil-Soluble Polymers and Specialty Polymers

Much research effort has been expended in the development of polymers that are more resistant to lubricating oils than polystyrene or natural rubber, and these efforts have been crowned with success. However, polymers also play a role in improving the viscosity of lubricating oils, improving pour depressant properties, and enhancing the rate of flow of crude oil in pipelines.

Atactic polypropylene and polyisobutylene were used as viscosity improvers (VI) in lubricating oils before many industrial scientists understood their effect in these oils. The first requirement is that the polymer be compatible with the oil at operating temperatures. When the solubility parameters are within ± 1.8 H, the polymer chains will tend to be fully extended and will serve as viscosity enhancers. The second requirement is that the polymer be incompatible with the oil at ordinary temperatures so that it will be present in a coiled conformation and have little effect on the viscosity of the oil.

These criteria would be difficult to meet without use of the Hildebrand solubility parameter expression and the Arrhenius equation, which show the effect of temperature on the solubility and viscosity (η) of the solutions, respectively. These expressions are shown below.

$$\delta = \left(\frac{D(\Delta H - RT)}{M}\right)^{1/2}$$

Hildebrand equation, showing change of δ with change in temperature

$$\eta = Ae\frac{E}{RT}$$

Arrhenius equation, showing a decrease in viscosity with temperature

The use of polyisobutyl methacrylate and polycyclohexylstyrene as viscosity improvers in oils was discovered through the use of these concepts. The formulas for the repeating units in these hot-oil-soluble polymers are shown below.

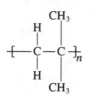

Polyisobutylene

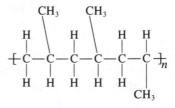

Atactic polypropylene

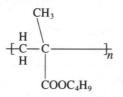

Polyisobutyl methacrylate

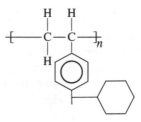

Polycyclohexylstyrene

Paraffin waxes are semicrystalline and hence tend to precipitate and impede liquid flow of crude oil. The addition of 1000 ppm ethylene-vinyl acetate copolymers with solubility parameters (δ = 8.6 H) similar to those of crude oil (δ = 8.9 H) reduces the tendency for the formation of incompatible paraffin waxes and serves as pour-point depressants. Similar improvements were shown when chlorinated polyethylene was used as a pour-point depressant. The formula for a repeating unit of the random copolymer of ethylene and vinyl acetate is shown below.

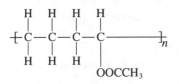

Low-molecular-weight oil-soluble polymers are also used as lubricating oils.

8.7 Applications of Polymers in Medicine

Because of the great amount of publicity about it, most people are aware of the polyurethane "artificial heart." They may also be aware of the use of plasticized PVC and polyethylene as catheters and tubing and of silicones for the reconstruction of body parts. Equally important is the use of HDPE-metal combinations for joints, polyethylene terephthalate (PET) mesh for blood vessel replacement, cellulose acetate membrane for external hemodialysis systems, poly-2-hydroxyethyl methacrylate for contact lenses, polymethyl methacrylate for intraocular lens implants, polytetrafluoroethylene (PTFE) buckles for the repair of detached retinas and for padding between joints, and filled polymethyl methacrylate for dentures.

8.8 Water-Soluble Polymers

Since water-soluble starch, guar gum, and other such materials have been available for centuries, the concept of synthetic water-soluble polymers is readily acceptable. The difference in water solubility between starch and cellulose is based on crystallinity and the different type of intermolecular hydrogen bonds present. These bonds in starch are primarily intramolecular and the hydrogen bonds in cellulose are intermolecular.

Thus, any chemical reaction that reduces the intermolecular bond strength and crystallinity promotes water solubility of cellulose. This objective was accomplished by Cross, Bevan, and Beadle in the early 1900s when they converted at least one of the three hydroxyl groups in the repeating unit of cellulose to xanthate by the reaction of carbon bisulfide with alkaline cellulose.

Many cellulose ethers with a degree of substitution (DS) of less than three are water-soluble. Among these are methylcellulose, hydroxyethylcellulose, hydroxypropylcellulose, and carboxymethylcellulose (CMC). Methylcellulose and hydroxypropylcellulose are soluble in cold water but form gels in hot water.

The most important water-soluble cellulose derivative is CMC, which is produced by the condensation of monochloroacetic acid with alkalicellulose.

CMC is available in DS ranges of 0.4 to 1.4. Because of the presence of the carboxyl group, its maximum viscosity occurs at a pH of 7 to 9. These water-soluble derivatives of cellulose are used as thickeners in food, as pharmaceuticals, for enhanced oil recovery, drilling fluids, gels, and for liquid-absorbents, drag reduction, and superabsorbents.

The hydrolyzed copolymer of styrene and maleic anhydride, which was patented in the 1920s, continues to be used as a textile assistant (Stymer) and a viscosity improver in waterborne paints. The hydrolyzed copolymer of vinyl methyl ether and maleic anhydride (Gantrez) has been used in furniture and floor polishes. Polyvinyl methyl ether is soluble in water, but other polyvinyl alkyl ethers are insoluble in water.

Polyvinyl alcohol (PVA) is an important water-soluble polymer produced by the hydrolysis of polyvinyl acetate. Since the monomer, vinyl alcohol, does not exist, this synthesis is of considerable scientific interest. PVA has been extruded as clear fibers (Kuralon) and film. These are insolubilized by reacting the surface hydroxyl groups with formaldehyde to produce polyvinyl formal.

PVA has excellent adhesion to cellulose and is used as a binder in cement formulation. The addition of as little as 0.1% borax ($Na_2B_4O_7$) to PVA causes gelation of aqueous solutions of PVA. PVA films have excellent resistance to permeation by gases. The equations for the production of PVA from PVAC are shown below.

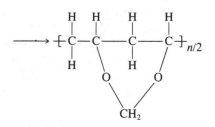

PVAC   **PVA**   **Formaldehyde**

Polyvinyl formal

Poly-*N*-vinyl-2-pyrrolidone (PVP, Periston) was produced in Germany during World War II and used as an extender for blood plasma. PVP and its water-soluble copolymers are also used as textile sizes and adhesives and for enhanced oil recovery. The formula for the repeating unit in PVP is shown below.

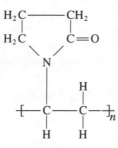

Polyacrylamide and its copolymers are used as flocculants in water and sewage treatment plants, as boiler scale preventives, and in enhanced oil recovery. The addition of small amounts of these polymers, under pressure, to water in oil wells increases the yield of oil. Similarly, improved flow is obtained when these polymers are added to water used for fire fighting. The formula for the repeating units of polyacrylamide is shown below.

8.9 Anaerobic Adhesives

Some acrylic monomers, such as tetramethylene glycol dimethacrylate, will not polymerize in the presence of oxygen, even when peroxy catalysts and accelerators are present. These monomers are stored in the presence of oxygen, and the oxygen is removed at the site of the application in which these adhesives are to be used, such as for setting bolts in concrete.

8.10 Ablative Polymers

Several polymer systems have been developed to combat the heat generated by passage of aerospace vehicles through the earth's atmosphere at high velocity. Polytetrafluoroethylene was used as the heat shield for the Venus entry probe; however, the formation of a porous solid is essential for other ablative materials. Carbon has the highest "heat of ablation" of any other material, providing that the residual particulates are not removed. The highest yield of carbon ablatives is produced by the pyrolysis of phenolic resins, and the particulate char can be stabilized by reinforcing the phenolic resin with graphite fibers.

Fiber-reinforced polymeric ablators act as heat sinks when ablated. Other polymers with high char yield are polyimides, polybenzimidazoles (PBI), polyphenylene oxide, polyphenylenes, and polymers of *p*-phenylphenols. A silica-fiber-reinforced epoxy novolac resin containing phenolic microballoons was used in the heat shield of the Apollo space vehicle.

8.11 Heat-Resistant Plastics for Aerospace Applications

Because of the need for heat-resistant polymers, the U.S. Air Force funded research resulting in the development of several temperature-resistant plastics, such as polybenzimidazole (PBI), which is being produced commercially as a high-performance fiber. The best heat resistance has been observed with aromatic and heterocyclic polymers with a ladder structure (that is, a double polymer chain) so that if one part of the chain is cleaved, the polymer's integrity is maintained. The formulas for the repeating unit of some of these heat-resistant polymers are shown below.

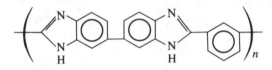

PBI

Polyphenylene

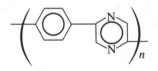

Polypyrazine

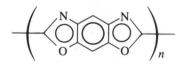

Polybenzoxazole

Polyquinoxaline (a ladder polymer)

8.12 The Future

As the polymer age matures, much of the new growth will be based on new applications of existing polymers and blends. Because of their versatility, these polymers will be used to solve many unsolved problems and to meet the requirements for many new applications.

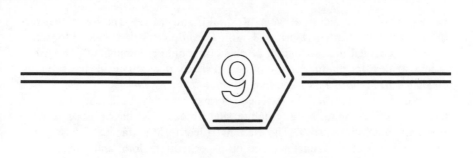

Moderately High-Performance Polymers

While the thermosets qualify as moderately high-temperature polymers, the general-purpose thermoplastics, which were available prior to World War II, would not withstand temperatures above 90 °C. This deficiency was overcome by the development of a terpolymer of styrene, maleic anhydride, and acrylonitrile, and a terpolymer of styrene, fumaronitrile, and polydichlorostyrene. Because of the toxicity of fumaronitrile and the high cost of polydichlorostyrene, the terpolymer of styrene, maleic anhydride, and acrylonitrile (Cadon) was marketable, and the use of this polymer is increasing.

9.1 Thermosets

We provided some information on phenolic, amino, and unsaturated polyesters in Chapter 4. Additional information on these thermosetting polymers will be presented in this chapter.

9.1.1 Phenolic Resins

Resins, produced by the condensation of phenol and formaldehyde, are used in the United States at an annual rate of over 1.2 million tons. Phenolic resins (PF) are the leading thermosets, and their principal use is in plywood (580,000 tons). About 100,000 tons of PF is used for molding compounds.

Phenolic resin molding compounds may be compression or transfer molded. They are produced by Occidental (Durez), Plastics Engineering, Fiberite, Reichhold Chemicals, Rogers, Resinoid, and Valite. The traditional producers, namely Borden (Durite), Union Carbide (Bakelite), and Monsanto, have discontinued the production of phenolic molding compounds. The use of asbestos fillers for the production of heat-resistant phenolic resins has been discontinued in favor of less hazardous fillers.

Wood flour, cotton flock, twisted cotton cord, rubber, diatomaceous silica, talc, clay, wollastonite, mica, graphite, and fibrous glass have all been used as fillers for phenolic plastics. The National Electric Manufacturers Association (NEMA) has graded laminates based on paper, cotton cloth, and glass cloth.

Both the phenol and the formaldehyde used to produce phenolic resins are toxic. However, cured phenolic resins are used in beer and soft drink can coatings, and no detrimental effects of phenolic resins have been reported. The properties of wood-flour-filled, mineral-filled, and fibrous-glass-filled phenolic plastics are presented in Table 9–1.

Table 9–1. Thermal, Physical, and Chemical Properties of Typical Phenolic Plastics

Property	Wood-flour-filled(a)	Mineral-filled	Glass-reinforced
Heat deflection temperature at 1820 kPa, °C	165	200	250
Maximum resistance to continuous heat, °C	160	175	175
Coefficient of linear expansion, cm/cm/°C $\times$ 10^{-5}	3.0	2.0	1.5
Compressive strength, kPa	172,400	172,400	120,000
Impact strength, Izod: cm · N/cm of notch	21.5	21.5	75
Tensile strength, kPa	48,250	41,400	60,000
Elongation, %	0.5	0.5	0.2
Hardness, Rockwell	M100	M110	E70
Specific gravity	1.4	1.5	1.85
Water absorption, %	0.4	0.03	0.5
Dielectric constant	6	8	5
Resistance to chemicals at 25 °C:(b)			
Nonoxidizing acids (20% H_2SO_4)	S	S	S
Oxidizing acids (10% HNO_3)	Q	Q	Q
Aqueous salt solutions (NaCl)	S	S	S
Aqueous alkaline solutions (NaOH)	Q	Q	Q
Polar solvents (C_2H_5OH)	S	S	S
Nonpolar solvents (C_6H_6)	S	S	S
Water	S	S	S

(a) Conversion tables appear in Appendix. (b) S, satisfactory; Q, questionable; U, unsatisfactory.

9.1.2 Furan Plastics

Since furan plastics are made from furfural, which is obtained by the hydrolysis of oat hulls or corn cobs, they have been of interest to those concerned with chemurgy. Furan resins were patented in 1928. Commercial furan resins are produced by the condensation of furfuryl alcohol with formaldehyde under acidic conditions. Carbon-flour-filled furan resins are used as molding powders. These dark resins are also used as plastic concrete and as alkali-acid reactant mortars.

Furan mortar (Alkor) is used for joining brick used as a sheathing in pickling vats and chemical processing equipment. The liquid furan resins have been used to impregnate plaster of Paris.

The formula of a repeating unit of a furan resin is shown below, and the properties of a carbon-filled furan plastic are presented in Table 9–2.

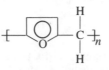

Table 9–2. Thermal, Physical, and Chemical Properties of a Typical Carbon-Filled Furan Resin(a)

Heat deflection temperature
at 1820 kPa, °C 80
Maximum resistance to continuous
heat, °C 100
Coefficient of linear expansion,
cm/cm/°C × 10^{-5} 7.5
Compressive strength, kPa 69,000
Flexural strength, kPa 34,475
Impact strength, Izod:
cm · N/cm of notch 21.4
Tensile strength, kPa 41,370
Elongation, % 1.5
Hardness, Rockwell R110
Specific gravity 1.7

Resistance to chemicals at 25 °C:(b)
Nonoxidizing acids (20% H_2SO_4) .... S
Oxidizing acids (10% HO_3) .... U
Aqueous salt solutions (NaCl) .... S
Aqueous alkaline solutions (NaOH) .... S
Polar solvents (C_2H_5OH) .... S
Nonpolar solvents (C_6H_6) .... S
Water .... S

(a) Conversion tables appear in Appendix. (b) S, satisfactory; Q, questionable; U, unsatisfactory.

9.1.3 Allylic Resins

Light-colored molding compounds are prepared by blending a prepolymer of diallyl phthalate (DAP) with filler, peroxy catalyst, and diallyl phthalate monomers. The fibrous-glass-filled molding products can be compression or transfer molded. The catalyzed prepolymer may also be used to impregnate glass cloth for the production of laminates. DAP is used as insulation, electrical and electronic parts, and as corrosion-resistant parts for chemical and food processing equipment. The formula for diallyl phthalate is shown below, and the properties of a glass-filled molded DAP are presented in Table 9–3.

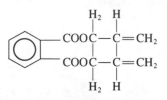

Table 9–3. Thermal, Physical, and Chemical Properties of a Typical Fibrous-Glass-Filled Allylic Plastic (DAP)(a)

Heat deflection temperature
   at 1820 kPa, °C............................... 200
Maximum resistance to continuous
   heat, °C.. 150
Coefficient of linear expansion,
   cm/cm/°C $\times$ 10^{-5} 2.0
Compressive strength, kPa................ 186,000
Flexural strength, kPa 131,000
Impact strength, Izod:
   cm · N/cm of notch 106
Tensile strength, kPa........................ 58,000
Elongation, % 4
Hardness, Rockwell............................. E80
Specific gravity 1.7
Water absorption.................................. 0.14
Dielectric constant.................................. 4

Resistance to chemicals at 25 °C:(b)
   Nonoxidizing acids (20% H_2SO_4) .... S
   Oxidizing acids (10% HNO_3) .... U
   Aqueous salt solutions (NaCl)................. S
   Aqueous alkalies (NaOH)...................... Q
   Polar solvents (C_2H_5OH) .... S
   Nonpolar solvents (C_6H_6) .... U
   Water .... S

(a) Conversion tables appear in Appendix. (b) S, satisfactory; Q, questionable; U, unsatisfactory.

Another commercially available allylic resin is diethylene glycol bis (allyl carbonate) (CR39). This monomer has been polymerized *in situ* to produce a scratch-resistant clear plastic with good optical properties. The formula for this monomer is shown below.

$$H_2C = CH — CH_2 — OOCOCH_2 — CH_2 — O — CH_2 CH_2OCOOCH = CH_2$$

Triallyl cyanurate (TAC) has been used as a heat-resistant cross-linking agent in polyester resins. The formula for TAC is shown below.

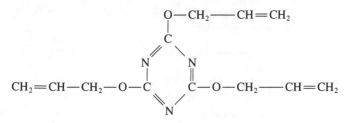

9.1.4 Alkyds

In addition to their uses in coatings (as described in Section 7.3), alkyds with comparable formulations are also used as compression and transfer molding powders. Alkyds, molded from molding-grade resins, are used in electrical and electronic applications. The properties of a typical fibrous-glass-filled alkyd molding are presented in Table 9–4.

9.1.5 Epoxy Plastics

In addition to their use as protective coatings and adhesives (as discussed in Section 7.6), epoxy resins (EP) are also used as molding powders. These resins may be reinforced with fibrous glass, or they may be filled with hollow glass spheres to produce syntactic plastics.

Epoxy resins (EP), which were originally used as structural adhesives for bonding metals in aircraft during World War II, are now used as protective coatings, fiber-reinforced laminates, concrete patches, and molded plastics. The prepolymer in adhesives, coatings, and laminates is cured (cross-linked) by the reaction of the epoxy groups with polyamines. The prepolymer in the molded plastics is cured by the reaction of anhydrides of polycarboxylic acids with the hydroxyl pendant groups on the epoxy polymer chain. The properties of molded EP are presented in Table 9–5.

9.2 Thermoplastics

9.2.1 Styrene–Maleic Anhydride Copolymers

Since styrene–maleic anhydride copolymers have good high-temperature properties but are difficult to mold, and copolymers of styrene and acrylonitrile

**Table 9–4. Thermal, Physical, and Chemical Properties of
Typical Alkyds(a)**

Heat deflection temperature
  at 1820 kPa, °C 200
Maximum resistance to continuous
  heat, °C.. 200
Coefficient of linear expansion,
  cm/cm/°C × 10^{-5} 2.0
Compressive strength, kPa 137,900
Flexural strength, kPa 103,425
Impact strength, Izod:
  cm · N/cm of notch 106.7
Tensile strength, kPa........................... 41,370
Elongation, % 2
Hardness, Rockwell 80E
Specific gravity 2.1

Resistance to chemicals at 25 °C:(b)
  Nonoxidizing acids (20% H_2SO_4) .... S
  Oxidizing acids (10% HNO_3) .... U
  Aqueous salt solutions (NaCl)................... S
  Aqueous alkalies (NaOH) .... S
  Polar solvents (C_2H_5OH) .... S
  Nonpolar solvents (C_6H_6) .... Q
  Water .... S

(a) Conversion tables appear in Appendix. (b) S,
satisfactory; Q, questionable; U, unsatisfactory.

have good resistance to impact and are readily moldable, it was logical to obtain a moldable, heat-resistant terpolymer by using all these monomers. The terpolymer (Cadon) was patented by Seymour in the 1940s but was not marketed until the late 1970s. It is now one of the fastest growing of the moderately high-performance plastics.

Although styrene tends to form alternating copolymers with maleic anhydride and acrylonitrile, this tendency may be overcome if the monomers are polymerized at temperatures above 105 °C. Random copolymers are produced at this elevated temperature.

The flexibility of styrene–maleic anhydride–acrylonitrile terpolymer has been increased by the substitution of ethyl acrylate for part of the styrene and by the addition of elastomers to these terpolymers.

The properties of typical styrene–maleic anhydride–acrylonitrile terpolymers are presented in Table 9–6.

9.2.2 Acrylonitrile-Butadiene-Styrene Copolymers

The first high-impact polystyrene (HIPS) was a blend of polystyrene and styrene-butadiene copolymers (SB), which was developed by Seymour in the

Table 9–5. Thermal, Physical, and Chemical Properties of Typical Molded Epoxy Plastics (EP)

Property	Epoxy plastic(a)	Glass-filled EP	Glass-sphere-filled EP
Heat deflection temperature at 1820 kPa, °C	140	150	115
Maximum resistance to continuous heat, °C	120	135	110
Coefficient of linear expansion, cm/cm/°C × 10^{-5}	2.5	2.0	2.5
Compressive strength, kPa	120,000	206,850	82,740
Flexural strength, kPa	124,100	103,400	41,370
Impact strength, Izod: cm · N/cm of notch	53.4	53.4	10.6
Tensile strength, kPa	51,710	82,740	41,370
Elongation, %	5	4	1
Hardness, Rockwell	M90	M105	–
Specific gravity	1.2	1.8	0.8
Dielectric constant	4	4	4
Water absorption	0.2	0.1	0.1
Resistance to chemicals at 25 °C:(b)			
Nonoxidizing acids (20% H_2SO_4)	S	S	S
Oxidizing acids (10% HNO_3)	U	U	U
Aqueous salt solutions (NaCl)	S	S	S
Aqueous alkalies (NaOH)	S	S	S
Polar solvents (C_2H_5OH)	S	S	S
Nonpolar solvents (C_6H_6)	S	S	S
Water	S	S	S

(a) Conversion tables appear in Appendix. (b) S, satisfactory; Q, questionable; U, unsatisfactory.

1940s. This flexibilizing technique was also used to improve the impact resistance of the styrene-acrylonitrile (30) copolymers (SAN). The elastomer used in the original blend was acrylonitrile rubber (NBR). The ratio of SAN to NBR was 64 to 36, but the ratio varies in many commercial compositions.

Borg-Warner's first blends were of cyclized rubber and butyl rubber, and this led to ABS resins, which were called Cycolac. Blends of SAN and NBR are still available, but much of today's ABS is made by grafting techniques. ABS plastics are also produced by Monsanto and Dow. These plastics have been marketed under the trade names Abson, Blendex, and Kralastic. ABS is used for electrical and electronic applications, pipe, automotive parts, and luggage. The latter is made by thermoforming ABS sheets. ABS is also blended with other polymers to increase their impact resistance. The properties of ABS plastics are presented in Table 9–7.

Table 9–6. Thermal, Physical, and Chemical Properties of a Typical Terpolymer of Styrene, Maleic Anhydride, and Acrylonitrile

Property	Unfilled plastic(a)	Fibrous-glass-reinforced plastic
Heat deflection temperature at 1820 kPa, °C	110	120
Maximum resistance to continuous heat, °C	100	110
Coefficient of linear expansion, cm/cm/°C $\times 10^{-5}$	8	6
Compressive strength, kPa	100,000	150,000
Flexural strength, kPa	100,000	150,000
Impact strength, Izod: cm $\cdot$ N/cm of notch	40	96
Tensile strength, kPa	55,000	80,000
Elongation, %	204	212
Hardness, Rockwell	R106	R110
Specific gravity	1.05	1.10
Water absorption, %	0.1	0.1
Resistance to chemicals at 25 °C:(b)		
Nonoxidizing acids (20% H_2SO_4)	S	S
Oxidizing acids (10% HNO_3)	Q	Q
Aqueous salt solutions (NaCl)	S	S
Aqueous alkalies (NaOH)	Q	Q
Polar solvents (C_2H_5OH)	S	S
Nonpolar solvents (C_6H_6)	U	U
Water	S	S

(a) Conversion tables appear in Appendix. (b) S, satisfactory; Q, questionable; U, unsatisfactory.

The resistance of ABS to outdoor service has been improved by substituting ethyl acrylate for butadiene in the terpolymer. The clarity of ABS has been improved by the substitution of methyl methacrylate for styrene.

9.2.3 Filled Polyolefins

Polyolefins, such as polypropylene and polymethylpentene (TPX), have unusually good physical properties, but they do not meet the requirements of moderately high-performance polymers. However, as shown by data in Table 9–8, the addition of talc or fibrous glass increases the heat deflection temperature to above 100 °C, and hence these filled polyolefins can be classified as moderately high-performance plastics.

Table 9–7. Thermal, Physical, and Chemical Properties of Typical ABS Plastics

Property	Extrusion grade(a)	20% Glass-reinforced
Heat deflection temperature at 1820 kPa, °C	90	100
Maximum resistance to continuous heat, °C	80	90
Coefficient of linear expansion, cm/cm/°C $\times$ 10^{-5}	9.5	2.0
Compressive strength, kPa	48,265	96,530
Flexural strength, kPa	62,055	103,425
Impact strength, Izod: cm $\cdot$ N/cm of notch	320.3	53.4
Tensile strength, kPa	34,475	75,845
Elongation, %	60	5
Hardness, Rockwell	R60	M85
Specific gravity	1.03	1.2
Dielectric constant	0.25	0.4
Resistance to chemicals at 25 °C:(b)		
Nonoxidizing acids (20% H_2SO_4)	S	S
Oxidizing acids (10% HNO_3)	U	U
Aqueous salt solutions (NaCl)	S	S
Aqueous alkalies (NaOH)	S	S
Polar solvents (C_2H_5OH)	Q	Q
Nonpolar solvents (C_6H_6)	Q	Q

(a) Conversion tables appear in Appendix. (b) S, satisfactory; Q, questionable; U, unsatisfactory.

9.2.4 Polyfluorocarbons

Polytetrafluoroethylene (PTFE, Teflon) is intractable but can be formed into useful shape by cold pressing the finely divided polymer into a preform, which is then heated in a mold at temperatures greater than 330 °C. Polytetrafluoroethylene has excellent resistance to high temperatures and corrosives, a low coefficient of friction, and excellent electrical properties.

Polychlorotrifluoroethylene (PCTFE, Kel-F) is not quite as good in its resistance to heat and chemicals, but it is more readily molded. It is used for gaskets, valve seats, and observation windows.

Polyvinylidene fluoride (PVF_2, Kynar) is also not quite as resistant to heat and corrosives as PCTFE, but it can be molded and extruded. It is used as pipe, heat-shrinkable tubing, monofilaments, and tank linings. Polyvinyl fluoride (PVF, Tedlar) is used primarily as a corrosion-resistant film.

Table 9–8. Thermal, Physical, and Chemical Properties of Typical Filled Polyolefins

Property	30% Glass-filled HDPE(a)	40% Talc-filled PP
Heat deflection temperature at 1820 kPa, °C	120	100
Maximum resistance to continuous heat, °C	110	120
Coefficient of linear expansion, cm/cm/°C $\times 10^{-5}$	5.0	6.5
Compressive strength, kPa	43,260	54,000
Flexural strength, kPa	75,800	58,600
Impact strength, Izod: cm · N/cm of notch	53.4	27
Tensile strength, kPa	62,000	30,000
Elongation, %	1.5	5
Hardness, Rockwell	R75	R95
Specific gravity	1.3	1.22
Water absorption, %	0.1	0.1
Dielectric constant	2	2
Resistance to chemicals at 25 °C:(b)		
Nonoxidized acids (20% H_2SO_4)	S	S
Oxidizing acids (10% HNO_3)	Q	Q
Aqueous salt solutions (NaCl)	S	S
Aqueous alkalies (NaOH)	S	S
Polar solvents (C_2H_5OH)	S	S
Nonpolar solvents (C_6H_6)	Q	Q
Water	S	S

(a) Conversion tables appear in Appendix. (b) S, satisfactory; Q, questionable; U, unsatisfactory.

Copolymers of polyethylene and fluorocarbon are used in applications in which the high-performance properties of PTFE are not essential. The coefficient of friction increases as the number of fluorine atoms per carbon atom decreases. The properties of these polyfluorocarbons are presented in Table 9–9.

Table 9–9. Thermal, Physical, and Chemical Properties of Typical Polyfluorocarbon Plastics

Property	PTFE(a)	PCTFE	PVDF	PVF	PE-CTFE	PE-TFE
Heat deflection temperature at 1820 kPa, °C	100	100	80	90	115	120
Maximum resistance to continuous heat, °C	250	200	150	125	100	160
Coefficient of linear expansion, cm/cm/°C $\times 10^{-5}$	10	14	8.5	10	8	7
Compressive strength, kPa	27,580	38,000	–	–	41,370	48,260
Flexural strength, kPa	–	60,000	–	–	48,260	38,000
Impact strength, Izod: cm · N/cm of notch	160	133.5	–	–	–	–
Tensile strength, kPa	24,130	34,475	55,160	–	48,260	48,260
Elongation, %	200	100	200	–	200	250
Hardness, Rockwell	D52	R80	R110	D64	R95	R50
Specific gravity	2.16	2.1	1.76	1.4	1.7	1.7
Dielectric constant	2	2.5	8	8	5	2.7
Water absorption, %	0	0	0	0	0	0
Resistance to chemicals at 25 °C:(b)						
Nonoxidizing acids (20% H_2SO_4)	S	S	S	S	S	S
Oxidizing acids (10% HNO_3)	S	S	S	Q	Q	Q
Aqueous salt solutions (NaCl)	S	S	S	S	S	S
Aqueous alkalies (NaOH)	S	S	S	S	S	S
Polar solvents (C_2H_5OH)	S	S	S	S	S	S
Nonpolar solvents (C_6H_6)	S	S	S	S	Q	Q
Water	S	S	S	S	S	S

(a) Conversion tables appear in Appendix. (b) S, satisfactory; Q, questionable; U, unsatisfactory.

Engineering Polymers

The original commercial polymers were used for applications such as billiard balls, fountain pens, and photographic film, for which metals were not suitable. These polymers and many other thermoplastics and thermosets, in a broad sense, are engineering polymers. However, in this chapter we will use a more restrictive and more practical definition related to properties, performance, and use criteria. High-performance polymers meeting these criteria are readily moldable, stiff thermoplastics with a good balance of tensile, compressive, impact, and shear strength, as well as good electrical properties and resistance to corrosives, and they retain these high-performance properties over a wide range of environmental applications. The Fisher-Pry theory has been used to show that high-performance plastics will replace 20 percent of all metals in the future. This is a 14-million-ton potential for new engineering materials. Allied, BASF, Bayer, Borg-Warner, Celanese, ICI, Mitsubishi, duPont, Polyplastics, and Asaki are the leading producers of these engineering polymers. Nylon, which was the first engineering thermoplastic, was introduced by duPont in the 1950s. These and other engineering polymers will be discussed in this chapter.

10.1 Acetals

Polymers of formaldehyde were investigated by Butlerov in 1859. For many decades, however, they were prevented from forming in aqueous formaldehyde solutions by the addition of methanol or by keeping the solution warm, and few attempts were made to use this polymer. In 1922, Staudinger re-examined this polymer but lost interest because of its thermal instability.

The instability problem was overcome by duPont chemists, who polymerized pure formaldehyde, produced by the decomposition of trioxane, in the presence of triphenylphosphine, and esterified the hydroxyl end groups in the product with acetyl groups. This high-molecular-weight crystalline polyoxymethylene (POM) is marketed under the trade name Delrin.

Hammerich and Boeree produced copolymers of formaldehyde in the 1960s. Walling, Brown, and Bartz recognized that unzipping of these copolymers would terminate when a stable carbon-carbon was reached in the thermal degradation. Hence, they patented stable copolymers of formaldehyde with small amounts of ethylene oxide. This copolymer was marketed by Celanese under the trade name Celcon.

POM plastics are characterized by good mechanical strength and rigidity, low coefficient of friction, lubricity, and excellent resistance to fatigue, impact, wear, and solvents. These polymers, usually injection molded, have unusually high shrinkage during molding, which must be compensated for in mold design. The commercial resins are widely used in the automobile industry for carburetors and fuel tanks, in construction for plumbing fixtures, in hardware as handles and portable tool components, in communications as telephone and stereo components, and in industrial applications as valves, cams, machine components, and pump impellers.

The unusually good thermal and physical properties of these acetal polymers are based on their crystallinity. Over 225,000 tons of polyacetals are produced annually worldwide. The annual production in the United States, Japan, and Western Europe is 80,000 tons, 80,000 tons, and 75,000 tons, respectively.

A summary of the properties of acetals is presented in Table 10–1, and the formulas for the repeating units are shown below.

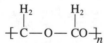

Acetal homopolymer   **Acetal copolymer**

10.2 Nylons

As described in Section 5.1, nylon-66 was developed by Carothers and used as a fiber in the late 1930s. However, since the hot melt spinning of this crystalline polymer was actually an extrusion process, it was apparent that this fiber-producing polymer could be used as a high-performance plastic. Nylon-66 has been used as an engineering plastic since the early 1950s.

Table 10–1. Thermal, Physical, and Chemical Properties of Acetal Polymers (POM)

Property	Homopolymer(a)	Copolymer	25% Glass-reinforced copolymer
Heat deflection temperature at 1820 kPa, °C	125	110	160
Maximum resistance to continuous heat, °C	100	100	125
Coefficient of linear expansion, $cm/cm/°C \times 10^{-5}$	10.0	8.5	5.0
Compressive strength, kPa	106,110	110,320	117,215
Flexural strength, kPa	96,530	89,635	193,060
Impact strength, Izod: cm · N/cm of notch	80.1	69.4	96.1
Tensile strength, kPa	68,950	62,055	128,600
Elongation, %	30	50	3
Hardness, Rockwell	M94	M78	M79
Specific gravity	1.412	1.41	1.61
Dielectric constant	3.2	3.7	4.0
Water absorption, %	0.25	0.25	0.3
Resistance to chemicals at 25 °C:(b)			
Nonoxidizing acids (20% H_2SO_4)	U	U	U
Oxidizing acids (10% HNO_3)	U	U	U
Aqueous salt solutions (NaCl)	S	S	S
Aqueous alkalies (C_2NOH)	S	S	S
Polar solvents (C_2H_5OH)	S	S	S
Nonpolar solvents (C_6H_6)	Q	Q	Q
Water	S	S	S

(a) Conversion tables appear in Appendix. (b) S, satisfactory; Q, questionable; U, unsatisfactory.

The crystallization of this diadic nylon may be controlled by nucleating, that is, seeding with nucleating agents, such as finely dispersed silica. Nylon-66 tends to degrade at temperatures above 65 °C, but this threshold may be increased to 140 °C with the addition of copper salts or phenolic derivatives.

All nylons are water-sensitive because of the presence of the hydrophilic amide group ($CONH_2$). Obviously, this sensitivity decreases as the ratio of methylene groups (CH_2) to amide groups increases. The absorbed water affects dimensional stability of molded parts but also serves as a volatile plasticizer. As is the case with other crystalline polymers, the heat deflection temperature of nylon-66 is increased and the coefficient of expansion decreased when the polymer is reinforced with additives, such as fibrous glass.

Recent investigations have revealed that the degradation of nylon at moderately high temperatures is the result of an oxidation of the amide groups, followed by the formation of a hydroperoxy group on the carbon atoms adjacent to the amide group. This series of oxidative steps is hindered by aromatic amines as well as by hindered phenolic and copper salts mentioned previously.

In addition to fibrous glass, kaolin, with a high aspect ratio, and mica have been used as reinforcements. The resistance of nylon to solvents has been improved by the addition of polyolefins.

The impact resistance of nylon has been improved by the addition of elastomers, ionomers, polyurethanes, and ABS terpolymers. In addition to being plasticized by the absorbed moisture, nylon is also plasticized by the addition of sulfonamides, polyols, and phthalic acid esters.

Because of their high strength, lubricity, and wear resistance, nylon-66 moldings and extrusions are used in bicycle parts, industrial machinery, mechanical drive components, bearings, chains, gears, pulleys, wire and cable, electric and electronic applications, and sprockets. The properties of nylons are shown in Table 10-2. Nylon-66 molding compounds are available from duPont (Zytel) and Monsanto (Vydyne).

Table 10–2. Thermal, Physical, and Chemical Properties of Nylon (PA)

Property	Nylon-66(a)	Glass-filled nylon-66	Nylon-6	Glass-filled nylon-6
Heat deflection temperature at 1820 kPa, °C	75	250	80	210
Maximum resistance to continuous heat, °C	120	140	125	130
Coefficient of linear expansion, cm/cm/°C $\times 10^{-5}$	8.0	2.0	8.0	3.0
Compressive strength, kPa	103,500	207,000	96,500	131,000
Flexural strength, kPa	103,500	276,000	96,500	207,000
Impact strength, Izod: cm·N/cm of notch	80	106.7	160	160
Tensile strength, kPa	82,750	172,000	62,055	172,000
Elongation, %	60	3	3	3
Hardness, Rockwell	R120	M100	M119	M101
Specific gravity	1.2	1.4	1.15	1.4
Dielectric constant	4.0	4.0	4.0	4.0
Water absorption, %	2.5	3.0	2.7	3.2
Resistance to chemicals at 25 °C:(b)				
Nonoxidizing acids (20% H_2SO_4)	U	U	U	U
Oxidizing acids (10% HNO_3)	U	U	U	U
Aqueous salt solutions (NaCl)	S	S	S	S
Aqueous alkalies (NaOH)	S	Q	S	Q
Polar solvents (C_2H_5OH)	Q	Q	Q	Q
Nonpolar solvents (C_6H_6)	S	S	S	S
Water	S	S	S	S

(a) Conversion tables appear in Appendix. (b) S, satisfactory; Q, questionable; U, unsatisfactory.

Schlenk obtained monadic nylon-6 by the polymerization of caprolactam in 1938, but this polymer was not used as a molding resin until the 1950s. Monsanto marketed an *in situ* polymerizable sodium caprolactamate, caprolactam, and initiator formulation for RIM nylon-6 (Nyrim) but has discontinued production of this polymer. However, Nylacast systems continues to market an RIM nylon-6 formulation. Nylon-6 molding compounds are available from Badische (Ultramid), Allied (Capron), and Mobay (Nydur). The product applications of nylon-6 are similar to those of nylon-66. The properties of nylon-6 are shown in Table 10–2.

Nylons with lower numbers, such as nylon-4 $(T_m, 265 \text{ °C})$, have higher water absorption and higher melting points than those with higher numbers, such as nylon-6,8 $(T_m, 240 \text{ °C})$, nylon-6,10 $(T_m, 225 \text{ °C})$, nylon-6,12 $(T_m, 212 \text{ °C})$, nylon-7 $(T_m, 223 \text{ °C})$, nylon-11 (Rilsan) $(T_m, 180 \text{ °C})$, and nylon-12 $(T_m, 180 \text{ °C})$. The water absorption of nylon-66 and nylon-6 is about 2.6%, whereas that of nylon-11 is 0.8%. Likewise, the dimensional stability change at 50% humidity decreases from 0.6 for nylon-66 and nylon-6 to 0.9 for nylon-11. The heat deflection temperature also decreases from about 80 °C for nylon-6 and nylon-612 to 55 °C for nylon-12.

Molded, amorphous nylon is produced by the condensation of dicarboxylic acids, such as adipic acid, with branched aliphatic amines, such as 2,2,4-trimethylhexamethylenediamine. In addition to the amorphous nylon (Trogamid T), flexibilized nylon copolymers (Vestamid) and polyether block amides (Pebax) are also available. The latter is produced by the condensation of a melt of a polyetherdiol and a polyamide with carboxylic acid end groups.

10.3 Polyimides (PI)

High-temperature-resistant polyimide film (duPont's H film) produced by the condensation of 4,4-diaminodiphenyl ether and pyromellitic anhydride with a heat deflection temperature of 280 °C has been available for a few decades. Thermoset polyimides have been fabricated by powder metallurgical techniques for use as jet engine parts and in the space shuttles.

An ordered PI resin (duPont, Vespel) can be solution spun from a methane sulfonic acid solution and used to reinforce epoxy resins for aircraft applications. About 2500 tons of polyimides are used annually in the United States. The formula for a typical repeating unit is shown below, and the properties of typical PI plastics are presented in Table 10–3.

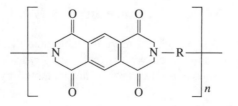

Table 10–3. Thermal, Physical, and Chemical Properties of Typical
Polyimides (PI)

Property	Thermoplastic(a)	Glass-filled thermoset (50%)
Heat deflection temperature at 1820 kPa, °C	315	350
Maximum resistance to continuous heat, °C	300	325
Coefficient of linear expansion, cm/cm/°C $\times$ 10^{-5}	5.0	1.3
Compressive strength, kPa	241,300	234,400
Flexural strength, kPa	172,400	144,800
Impact strength, Izod: cm $\cdot$ N/cm of notch	80	294
Tensile strength, kPa	96,500	44,000
Elongation, %	8	0.5
Hardness, Rockwell	E60	M118
Specific gravity	1.4	1.6
Dielectric constant	3.4	3.5
Water absorption, %	0	0.2
Resistance to chemicals at 25 °C:(b)		
Nonoxidizing acids (20% H_2SO_4)	Q	Q
Oxidizing acids (10% HNO_3)	Q	Q
Aqueous salt solutions (NaCl)	S	S
Aqueous alkalies (NaOH)	U	U
Polar solvents (C_2H_5OH)	S	S
Nonpolar solvents (C_6H_6)	S	S
Water	S	S

(a) Conversion tables appear in Appendix. (b) S, satisfactory;
Q, questionable; U, unsatisfactory.

PI films are resistant to exposure to thermal neutrons and other radiation except ultraviolet radiation. Most polyimides deteriorate when exposed to ultraviolet light. PI is insoluble in most solvents, but some special polyimides (Ciba-Geigy) are soluble in tetrahydrofuran. Polyimide moldings, composites, films, adhesives, fibers, and foams are used in electronic and aerospace applications, such as piston rings, rotary seal rings, wire insulation, and flame-resistant films, fabrics, and foams.

10.4 Polyamide Imides (PAI)

Polyamide imides (PAI), which are more tractable than PI, are produced when aromatic amines are condensed with trimellitic anhydride. These injection-moldable high-performance plastics, which are marketed by Amoco under the trade name Torlon, have been reinforced and used as materials of construction for an internal combustion automobile motor (Polimotor). This motor, which has been used in racing cars, is quieter and more fuel-efficient than conventional gasoline motors.

Some commercial polyamide imides contain about 0.5% PTFE and about 3% titanium oxide for lubricity and stability. PAI is used for high-temperature electrical connectors and other electrical, electronic, and aerospace applications.

The properties of PAI are shown in Table 10–4, and the repeating unit of a typical polyamide imide is shown below.

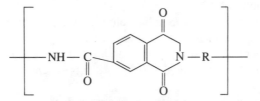

Table 10–4. Thermal, Physical, and Chemical Properties of a Typical Polyamide Imide (PAI)(a)

Heat deflection temperature at
  1820 kPa, °C 275
Maximum resistance to continuous
  heat, °C..225
Coefficient of linear expansion,
  cm/cm/°C × 10^{-5} 3.6
Compressive strength, kPa................. 220,000
Flexural strength, kPa 210,000
Impact strength, Izod:
  cm · N/cm of notch 133.5
Tensile strength, kPa........................ 186,000
Elongation, % 12
Hardness, Rockwell............................M119
Specific gravity 1.4
Dielectric constant................................4.0
Water absorption, % 0.2

Resistance to chemicals at 25 °C:(b)
  Nonoxidizing acids (20% H_2SO_4) .... S
  Oxidizing acids (10% HNO_3) ....U
  Aqueous salt solutions (NaCl)................... S
  Aqueous alkalies (NaOH).........................U
  Polar solvents (C_2H_5OH)..........................S
  Nonpolar solvents (C_6H_6) .... S
  Water .... S

(a) Conversion tables appear in Appendix. (b) S, satisfactory; Q, questionable; U, unsatisfactory.

10.5 Polyetherimide

The degree of chain flexibility of imide-type polymers may be increased by the insertion of *bis*-phenol-A units in the polymer chain. The resulting polyetherimides (PEI) are amorphous, amber yellow, transparent plastics, which were developed by Wirth in 1970 and are marketed by General Electric under the trade name Ultem.

PEI is resistant to most solvents but is soluble in some chlorinated aliphatic compounds. It is also resistant to ultraviolet and gamma radiation. It is transparent to microwaves and has good resistance to flame, as demonstrated by an oxygen index rating of 47%.

PEI behaves as a Hookean solid below its yield point. Its dielectric constant is essentially unchanged with increases in frequency up to 10^9 Hz. PEI retains its excellent physical properties at elevated temperatures. Underwriters' Laboratories (UL) has designated 170 °C as its upper temperature limit for continuous use.

PEI has been used in under-the-hood applications in automobiles, in appliances subjected to high temperatures, and in integrated circuits, switches, printed wiring cords, aerospace applications, and wire insulation.

The formula for the repeating unit of PEI is shown below, and the properties of PEI are presented in Table 10–5.

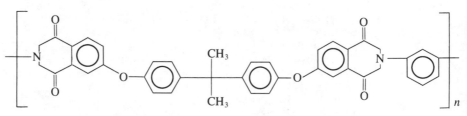

10.6 Polycarbonates

Polycarbonates (PC), the second most widely used engineering polymers, are produced by the condensation of phosgene ($COCl_2$) and *bis*-phenol-A. These polyesters were synthesized by Einhorn and Bischoff in 1898, by von Hedentron in 1922, and by Carothers in 1930. However, these polymers were not commercialized until the late 1950s when Bayer, General Electric, Mobay, and Taijen marketed these engineering polymers under the trade names Makrolon, Lexan, Merlon, and Panlite, respectively. Schnell of Farbenfabriken Bayer AG and Fox of General Electric introduced these important polymers independently.

The physical and thermal properties of these polymers are improved and mole shrinkage is reduced by reinforcement with fibrous glass. Since PC is sen-

Table 10–5. Thermal, Physical, and Chemical Properties of a Typical Polyetherimide (PEI)

Property	PEI(a)	10% Glass	20% Glass	30% Glass
Heat deflection temperature at 1820 kPa, °C	190	200	205	210
Maximum resistance to continuous heat, °C	170	175	180	185
Coefficient of linear expansion, cm/cm/°C × 10^{-5}	5.6	4.4	3.2	2.0
Compressive strength, kPa	140,000	155,000	169,000	176,000
Flexural strength, kPa	145,000	195,000	205,000	225,000
Impact strength, Izod: cm · N/cm of notch	133.5	146	213	267
Tensile strength, kPa	104,000	114,000	138,000	169,000
Elongation, %	6.0	6.0	3.0	3.0
Hardness, Rockwell	M109	M115	M120	M125
Specific gravity	1.27	1.35	1.45	1.6
Dielectric constant	3.1	3.3	3.5	3.7
Water absorption, %	0.06	0.1	0.15	0.2
Resistance to chemicals at 25 °C:(b)				
Nonoxidizing acids (20% H_2SO_4)	S	S	S	S
Oxidizing acids (10% HNO_3)	U	U	U	U
Aqueous salt solutions (NaCl)	S	S	S	S
Aqueous alkalies (NaOH)	Q	Q	Q	Q
Polar solvents (C_2H_5OH)	S	S	S	S
Nonpolar solvents (C_6H_6)	S	S	S	S
Water	S	S	S	S

(a) Conversion tables appear in Appendix. (b) S, satisfactory; Q, questionable; U, unsatisfactory.

sitive to moisture, it must be dried before processing. Under moderate conditions of humidity, PC behaves as a Hookean solid up to its yield point.

The creep of unfilled PC at 25 °C is negligible for loads not exceeding 14,000 kPa. It has excellent resistance to impact, but some deformation may occur under continuous impact. This clear plastic is stable to moderately strong corrosives, but highly stressed parts may be crazed if exposed to a hot, moist environment.

Polycarbonates are produced at an annual rate worldwide of 300,000 tons and in Japan, Western Europe, and the United States at an annual rate of 50,000, 75,000, and 140,000 tons, respectively. These polymers are characterized by good light transmission (90%), toughness, and resistance to moderately high temperatures. Polycarbonates are used for glazing, automotive applications, large containers, and ovenware. The formula for the repeating unit of polycarbonates is shown below, and their properties are presented in Table 10–6.

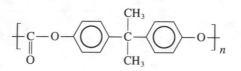

10.7 Aromatic Polyesters

Polyethylene terephthalate fibers (PET) were discussed in Section 5.4. PET is a crystalline polymer. Crystallization occurs more rapidly in lower-molecular-weight polymers and can be induced by the presence of nucleating agents. The amorphous state in these nonengineering polyesters is favored by quick quenching of film and molded parts. Thus, PET used for the blow molding of soft

Table 10–6. Thermal, Physical, and Chemical Properties of Typical Polycarbonates

Property	Unfilled(a)	20% Glass-filled
Heat deflection temperature, °C	130	145
Maximum resistance to continuous heat, °C	115	130
Coefficient of linear expansion, cm/cm/°C × 10^{-5}	6.8	2.2
Compressive strength, kPa	86,000	124,000
Flexural strength, kPa	93,000	158,000
Impact strength, Izod: cm · N/cm of notch	534	106
Tensile strength, kPa	72,000	131,000
Elongation, %	110	4
Hardness, Rockwell	M70	M92
Specific gravity	1.2	1.4
Dielectric constant	3	4
Water absorption, %	0.15	0.25
Resistance to chemicals at 25 °C:(b)		
Nonoxidizing acids (20% H_2SO_4)	Q	Q
Oxidizing acids (10% HNO_3)	U	U
Aqueous salt solutions (NaCl)	S	S
Aqueous alkalies (NaOH)	U	U
Polar solvents (C_2H_5OH)	S	S
Nonpolar solvents (C_6H_6)	U	U
Water	S	S

(a) Conversion tables appear in Appendix. (b) S, satisfactory; Q, questionable; U, unsatisfactory.

drink bottles is highly amorphous, but PET used for engineering applications is highly crystalline.

PET is today's major fiber and an important engineering polymer. Over 225,000 tons of PET is produced annually worldwide, and this volume is increasing. The United States is the leading user of aromatic polyester engineering resins (130,000 tons) and is followed by Japan and Western Europe with 25,000 and 75,000 tons, respectively. The catalyst for growth is the use of PET in soft drink bottles. This application was developed by N. C. Wyeth of duPont, whose principal contribution was orienting the PET in these bottles.

Molding-grade PET is produced by duPont (Rynite), Allied (Petra), and Mobay (Petlon). As shown in Table 10–7, the good properties of PET are enhanced by fibrous-glass reinforcement.

The processibility of aromatic polyesters is enhanced by the introduction of more methylene groups in the repeating units. Thus, as shown by the follow-

Table 10–7. Thermal, Physical, and Chemical Properties of Typical Polyethelene Terephthalate (PET)

Property	PET(a)	PET with 30% fibrous glass
Heat deflection temperature at 1820 kPa, °C	100	226
Maximum resistance to continuous heat, °C	100	160
Coefficient of linear expansion, cm/cm/°C $\times 10^{-5}$	6.5	2.9
Compressive strength, kPa	86,000	172,000
Flexural strength, kPa	112,300	234,000
Impact strength, Izod: cm $\cdot$ N/cm of notch	26.7	50
Tensile strength, kPa	62,000	158,000
Elongation, %	100	2.5
Hardness, Rockwell	M96	M100
Specific gravity	1.35	1.56
Dielectric constant	3.6	4.0
Water absorption, %	0.2	0.05
Resistance to chemicals at 25 °C:(b)		
Nonoxidizing acids (20% H_2SO_4)	S	S
Oxidizing acids (10% HNO_3)	Q	Q
Aqueous salt solutions (NaCl)	S	S
Aqueous alkalies (NaOH)	S	S
Polar solvents (C_2H_5OH)	S	S
Nonpolar solvents (C_6H_6)	S	S
Water	S	S

(a) Conversion tables appear in Appendix. (b) S, satisfactory; Q, questionable; U, unsatisfactory.

ing formulas, polybutylene terephthalate (PBT) has four methylene groups compared with two methylene groups in PET.

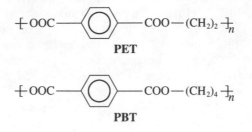

PET

PBT

The properties of PBT and PBT reinforced with 30% fibrous glass are presented in Table 10–8.

Table 10–8. Thermal, Physical, and Chemical Properties of Typical Polybutylene Terephthalate (PBT)

Property	Unfilled PBT(a)	PBT with 30% fibrous glass
Heat deflection temperature at 1820 kPa, °C	65	200
Maximum resistance to continuous heat, °C	60	150
Coefficient of linear expansion, cm/cm/°C $\times 10^{-5}$	7.0	2.5
Compressive strength, kPa	75,000	120,000
Flexural strength, kPa	96,000	110,000
Impact strength, Izod: cm · N/cm of notch	53.4	50
Tensile strength, kPa	55,000	117,000
Elongation, %	100	4
Hardness, Rockwell	M70	M90
Specific gravity	1.35	1.5
Dielectric constant	4.0	4.0
Water absorption, %	0.05	0.05
Resistance to chemicals at 25 °C:		
Nonoxidizing acids (20% H_2SO_4)	S	S
Oxidizing acids (10% HNO_3)	Q	Q
Aqueous salt solutions (NaCl)	S	S
Aqueous alkalies (NaOH)	S	S
Polar solvents (C_2H_5OH)	S	S
Nonpolar solvents (C_6H_6)	S	S
Water	S	S

(a) Conversion tables appear in Appendix. (b) S, satisfactory; Q, questionable; U, unsatisfactory.

An injection-moldable clear polyester (Kodel) is produced by Eastman by the condensation of 1,6-dihydroxycyclohexane and terephthalic acid. The formula for the repeating unit in Kodel is shown below.

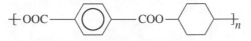

Polyarylates were produced independently by Conix, Levine, and Temin and Eareckson in the late 1950s. These aromatic esters are clear, high-temperature-resistant thermoplastics characterized by excellent resistance to ultraviolet radiation and to flame. These high-performance aromatic polyesters are produced by the condensation of *bis*-phenol-A with a mixture of iso and terephthalic acids. Polyarylates are produced in Japan by Itika, imported to the United States by Union Carbide, and marketed under the trade name Ardel. They are also produced and marketed by Celanese under the trade name Durel. Upon exposure to ultraviolet radiation, they undergo a photo-Fries rearrangement on the surface that produces 2-hydroxybenzophenones, which are ultraviolet light stabilizers.

Polyarylates are used in many outdoor applications, such as traffic lights, aircraft canopies, high-temperature lighting, and other electrical applications. Whether these high-temperature-resistant polymers are amorphous or crystalline depends on the reactants used. Crystallinity is favored when hydroquinone or more rigid diphenols are used in place of *bis*-phenol-A.

The amorphous *bis*-phenol-A based polyarylates are characterized by high glass transition temperatures ($T_g = 188\ °C$) and a low-temperature relaxation of $-75\ °C$, which contribute to a broad temperature range of good physical properties and toughness, respectively. Polyarylates have a limiting oxygen index of 38%, and a relatively low smoke density ($2.5\ N/cm^2$). Polyarylates have been used as ultraviolet stabilizers in other polymers.

The formula for the repeating unit is shown below, and the properties of polyarylates are presented in Table 10–9.

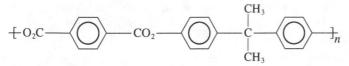

Another high-temperature-resistant polyester is produced by the condensation of *p*-hydroxybenzoic acid. This polymer, which is marketed under the trade name Ekonol, does not melt below its decomposition temperature of 450 °C. The formula of its repeating unit is shown below, and its properties are presented in Table 10–10.

$$\text{+O}\!\!-\!\!\bigcirc\!\!-\!\!CO_2 \text{+}_n$$

Table 10–9. Thermal, Physical, and Chemical Properties of Typical Polyarylates(a)

Heat deflection temperature at
  1820 kPa, °C 175
Maximum resistance to continuous
  heat, °C... 150
Coefficient of linear expansion,
  cm/cm/°C × 10^{-5} 6.5
Compressive strength, kPa 93,000
Flexural strength, kPa 79,800
Impact strength, Izod:
  cm · N/cm of notch 215
Tensile strength, kPa......................... 68,000
Elongation, % 50
Hardness, Rockwell R125
Specific gravity 1.2
Dielectric constant................................. 0.7
Water absorption, % 0.26

Resistance to chemicals at 25 °C:(b)
  Nonoxidizing acids (H_2SO_4).................... S
  Oxidizing acids (HNO_3)........................ Q
  Aqueous salt solutions (NaCl)................. S
  Aqueous alkalies (NaOH)....................... S
  Polar solvents (C_2H_5OH)........................ S
  Nonpolar solvents (C_6H_6) .... S
  Water .... S

(a) Conversion tables appear in Appendix. (b) S, satisfactory; Q, questionable; U, unsatisfactory.

10.8 Polyphenylene Oxide

Most engineering polymers are produced by unique condensation reactions. The reactant for the synthesis of polyphenylene oxide (PPO), which was discovered by Fox of General Electric, is truly unique. It involves a copper (I) amine-catalyzed oxidative coupling of 2,6-dimethylphenol. Because of the linear-bonded, sterically hindered aromatic ring structure, PPO is stiff, has a high melting point, and is difficult to process.

The processibility of this engineering polymer is improved, without significant loss in physical properties, by blending with polystyrene, which is compatible with PPO. In addition to the production of Noryl by General Electric, Borg-Warner is producing an oxidized, coupled copolymer of 2,6-dimethylphenol and 2,3,6-trimethylphenol (Prevex), which, like Noryl, is blended with polystyrene to improve processibility. Asaki is also producing PPO under the trade name Xyron.

PPO is a registered trademark of General Electric for poly 1,4- (2,6-dimethylphenyl) oxide. These polymers are characterized by excellent resistance

Table 10–10. Thermal, Physical, and Chemical Properties of Typical Crystalline Polyarylates (Ekonol)(a)

Heat deflection temperature, °C............... 300
Maximum resistance to continuous
   heat, °C.. 250
Coefficient of linear expansion,
   $cm/cm/°C \times 10^{-5}$ 71
Compressive strength, kPa.................265,000
Flexural strength, kPa 63,000
Impact strength, Izod:
   $cm \cdot N/cm$ of notch 25
Tensile strength, kPa......................... 60,000
Elongation, % 15
Hardness, Rockwell........................... R130
Specific gravity 1.44
Dielectric constant.................................. 3
Water absorption, % 0.02

Resistance to chemicals at 25 °C:(b)
   Nonoxidizing acids (20% H_2SO_4) .... S
   Oxidizing acids (10% HNO_3) .... Q
   Aqueous salt solutions (NaCl)................. S
   Aqueous alkalies (NaOH)...................... S
   Polar solvents (C_2H_5OH)........................ S
   Nonpolar solvents (C_6H_6) .... S
   Water .... S

(a) Conversion tables appear in Appendix. (b) S, satisfactory; Q, questionable; U, unsatisfactory.

to hydrolysis by acids, salts, alkalies, and hot water. However, PPO is attacked by many nonpolar solvents, such as amines, esters, halogenated compounds, and some hydrocarbons. These resins darken on exposure to ultraviolet radiation, but there is little noticeable loss in physical properties with exposure to ultraviolet, gamma, or X-ray radiation. Flame-resistant formulations are available.

Over 120,000 tons of PPO is produced annually worldwide. Of this, the United States, Western Europe, and Japan account for 68,000, 36,000, and 20,000 tons, respectively.

PPO is used for automotive parts, appliances, business machines, and electrical parts. The formula for the repeating unit of PPO is shown below, and its properties are presented in Table 10–11.

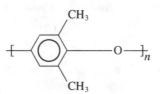

Table 10–11. Thermal, Physical, and Chemical Properties of a Polyphenylene Oxide (PPO)

Property	PPO(a)	Glass-filled PPO
Heat deflection temperature at 1820 kPa, °C	100	145
Maximum resistance to continuous heat, °C	80	130
Coefficient of linear expansion, cm/cm/°C $\times 10^{-5}$	5.0	2.0
Compressive strength, kPa	96,000	123,000
Flexural strength, kPa	89,000	144,000
Impact strength, Izod: cm · N/cm of notch	270	107
Tensile strength, kPa	55,000	120,000
Elongation, %	50	4
Hardness, Rockwell	R115	R115
Specific gravity	1.1	1.1
Dielectric constant	2.8	3.0
Water absorption, %	0.7	1.0
Resistance to chemicals at 25 °C:(b)		
Nonoxidizing acids (20% H_2SO_4)	S	S
Oxidizing acids (10% HNO_3)	Q	Q
Aqueous salt solutions (NaCl)	S	S
Aqueous alkalies (NaOH)	S	S
Polar solvents (C_2H_5OH)	S	S
Nonpolar solvents (C_6H_6)	U	U
Water	S	S

(a) Conversion tables appear in Appendix. (b) S, satisfactory; Q, questionable; U, unsatisfactory.

10.9 Polyphenylene Sulfide (PPS)

Polyphenylene sulfide (PPS) was synthesized by Genvresse in 1895 and by Lenz in the 1950s by the Friedel-Crafts and Ullman reactions, respectively.

Polyphenylene sulfide (PPS) was produced and patented by Edmonds and Hill in 1967. The patent was assigned to Phillips Petroleum, which markets this crystalline aromatic engineering polymer under the trade name Ryton.

PPS is characterized by a high melting point (235 °C) and excellent resistance to corrosives and solvents. It is unaffected by neutron or gamma radiation and has moderate resistance to ultraviolet radiation. This polymer cross-links and releases sulfur dioxide and carbonyl sulfide when heated at elevated temperatures. PPS has excellent flame retardancy and is rated V-0 by Underwriters' Laboratories.

PPS is used for automotive components, electrical appliances, and cookware. This polymer is produced by the Wurtz-Fittig condensation of *p*-dichlorobenzene and sodium sulfide. Kureha Chemical is producing PPS by a different polymerization method in Japan. Molded parts may be annealed by additional heating. The processibility of this polymer has been improved by the synthesis of polyphenylene ether sulfide imide. The formula for the repeating unit of PPS is shown below, and its properties are presented in Table 10–12.

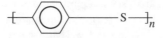

10.10 Polysulfones

Polysulfone was synthesized by Rose by the Friedel-Crafts self-condensation of diphenylene oxide sulfonyl chloride. This amorphous, transparent, temperature-resistant engineering polymer may also be produced by the condensation of 4,4-dichlorophenyl sulfone with an alkali salt of *bis*-phenol-A. These engineer-

Table 10–12. Thermal, Physical, and Chemical Properties of a Typical Polyphenylene Sulfide (PPS)

Property	Unfilled PPS(a)	40% Glass-filled PPS
Heat deflection temperature at 1820 kPa, °C	135	250
Maximum resistance to continuous heat, °C	110	200
Coefficient of linear expansion, cm/cm/°C $\times 10^{-5}$	5.0	2.2
Compressive strength, kPa	110,000	145,000
Flexural strength, kPa	96,000	207,000
Impact strength, Izod: cm $\cdot$ N/cm of notch	21	75
Tensile strength, kPa	74,000	141,000
Elongation, %	1.1	1
Hardness, Rockwell	R123	R123
Specific gravity	1.3	1.6
Dielectric constant	3.8	4.6
Water absorption, %	0.02	0.03
Resistance to chemicals at 25 °C:(b)		
Nonoxidizing acids (20% H_2SO_4)	S	S
Oxidizing acids (10% HNO_3)	S	S
Aqueous salt solutions (NaCl)	S	S
Aqueous alkalies (NaOH)	S	S
Polar solvents (C_2H_5OH)	S	S
Nonpolar solvents (C_6H_6)	S	S
Water	S	S

(a) Conversion tables appear in Appendix. (b) S, satisfactory; Q, questionable; U, unsatisfactory.

ing polymers are marketed by ICI (Victrex), Union Carbide (Udel), and 3M (Astrel). The inherent oxidative stability is related to the stiffening effect of the sulfonyl group, and the flexibility is related to the presence of ether groups in the polymer chain.

Polyarylsulfones are characterized by good resistance to environmental stress cracking and to hydrolysis by acids and bases over a wide range of temperatures. Polyether sulfone (PES) is a transparent, amorphous engineering polymer that is more flexible than polyarylsulfone (PES), which is characterized by excellent creep values, low flammability (LOI, 40%), and low smoke emission. Polysulfone is transparent and light amber in color. Solutions of polysulfone in water-soluble organic compounds are used to produce hollow-fiber membranes used for gas separations.

The total annual consumption of polysulfones in the United States is about 5000 tons. These polymers are used for pipe, flow meters, fuel cells, membranes, microwave cookware, and electrical applications.

Equations for the production of polysulfones are shown below, and the properties of these polymers are presented in Table 10–13.

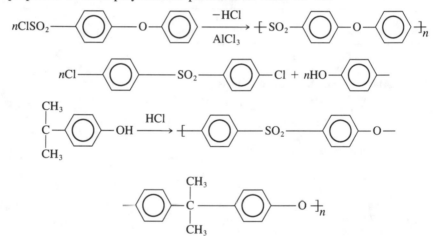

10.11 Polyether Ether Ketone

Like the polysulfones, polyether ether ketone (PEEK) was also invented by Rose in the 1970s. This crystalline engineering polymer is marketed by ICI under the trade name Victrex. The carbonyl and ether groups in this wholly aromatic polymer chain serve as stiffening and flexibilizing groups, respectively. PEEK is characterized by a high melting point (330 °C) and a high degree of oxidative stability. It has a limiting oxygen index (LOI) of 34 to 35 and an Underwriters' Laboratories rating of V-0. PEEK is being used for military and nuclear applications and for compressor parts.

Table 10–13. Thermal, Physical, and Chemical Properties of Typical Polysulfones

Property	Polysulfone(a)	Polyether sulfone
Heat deflection temperature at 1820 kPa, °C	175	205
Maximum resistance to continuous heat, °C	150	165
Coefficient of linear expansion, cm/cm/°C $\times 10^{-5}$	5.4	5.5
Compressive strength, kPa	96,000	96,000
Flexural strength, kPa	107,000	127,000
Impact strength, Izod: cm · N/cm of notch	80	80
Tensile strength, kPa	82,000	82,000
Elongation, %	25	25
Hardness, Rockwell	M69	M88
Specific gravity	1.24	1.37
Dielectric constant	3.1	3.1
Water absorption, %	0.3	0.4
Resistance to chemicals at 25 °C:(b)		
Nonoxidizing acids (20% H_2SO_4)	S	S
Oxidizing acids (10% HNO_3)	U	U
Aqueous salt solutions (NaCl)	S	S
Aqueous alkalies (NaOH)	S	S
Polar solvents (C_2H_5OH)	S	S
Nonpolar solvents (C_6H_6)	Q	Q
Water	S	S

(a) Conversion tables appear in Appendix. (b) S, satisfactory; Q, questionable; U, unsatisfactory.

The formula for the repeating unit of PEEK is shown below, and its properties are presented in Table 10–14.

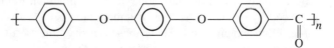

10.12 Liquid-Crystal Polymers

Since polymers exist as long chains, they have potential as liquid crystals. However, this potential is seldom realized because of their intractability below the decomposition temperatures. It has been recognized that petroleum pitches, poly-*p*-xylylene, polysiloxane, polyether, and polyesters exhibit anisotropic, liquid-crystal properties in fluid melts and that these conformations could be "locked in" by the incorporation of mesogenic entities in these polymer chains.

Table 10–14. Thermal, Physical, and Chemical Properties of a Typical Polyether Ether Ketone (PEEK)

Property	Unfilled PEEK(a)	40% Glass-filled PEEK
Heat deflection temperature at 1820 kPa, °C	150	300
Maximum resistance to continuous heat, °C	125	225
Coefficient of linear expansion, cm/cm/°C $\times 10^{-5}$	5.5	2.2
Compressive strength, kPa	90,000	125,000
Flexural strength, kPa	110,000	250,000
Impact strength, Izod: cm $\cdot$ N/cm of notch	50	75
Tensile strength, kPa	70,000	107,000
Elongation, %	50	2
Hardness, Rockwell	R123	R123
Specific gravity	1.3	1.5
Dielectric constant	3.2	3.5
Water absorption, %	0.15	0.12
Resistance to chemicals at 25 °C:(b)		
Nonoxidizing acids (20% H_2SO_4)	S	S
Oxidizing acids (10% HNO_3)	S	S
Aqueous salt solutions (NaCl)	S	S
Aqueous alkalies (NaOH)	S	S
Polar solvents (C_2H_5OH)	S	S
Nonpolar solvents (C_6H_6)	S	S
Water	S	S

(a) Conversion tables appear in Appendix. (b) S, satisfactory; Q, questionable; U, unsatisfactory.

Several aromatic polyesters, such as those described in Section 10.10, meet the specifications for liquid crystals. One of the commercial liquid-crystal polymers is produced by the condensation of terephthalic acid, p-hydroxybenzoic acid, and p,p-dihydroxybiphenyl. These low-viscosity, injection-moldable, heat-resistant polymers undergo parallel ordering in the molten state, which results in tightly packed fibrous chains in molded parts.

These liquid-crystal polymers are nematic self-reinforcing polymers. These anisotropic molded specimens retain their outstanding physical properties at high temperatures. The tensile strength of these engineering polymers is 70,000 kPa at 300 °C, and these properties improve as the molding is cooled to subzero temperatures.

These engineering polymers have an Underwriters' Laboratories rating of 240 °C for continuous electrical service and can withstand intermittent temperatures as high as 315 °C. Commercial liquid polymers have good resistance to flame, as indicated by an LOI of 42 and a V-0 UL 94 procedure rating. The smoke generation, as measured by U.S. National Bureau of Standards (NBS) smoke chamber tests, is low (7.5 to 12 cm per 4 min.); actually, they are intumescent when burning. These aromatic copolyesters are resistant to most solvents and corrosives but are attacked by boiling aqueous caustic solutions. They are transparent to microwaves and are unaffected by ultraviolet and ionizing radiation.

Liquid-crystal polymers provide a wide range of properties. They are capable of replacing metals and ceramics in many applications. These wholly aromatic engineering polymers are marketed under the trade names Xydar by Dartco, Calendar by Celanese, and Polyester X7G by Eastman. Moldings of these rigid, rodlike, heat-resistant engineering polymers may be used in place of metals and ceramics for electronics, aerospace, and transportation applications.

The formula for a repeating unit of filled and unfilled liquid-crystal polymers is shown below, and their properties are presented in Table 10–15.

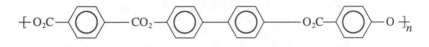

10.13 Polymer Blends

Blending of fibers, elastomers, and plastics has been practiced by the polymer industry for several decades. The textile designer is well aware of the advantage of blending hydrophobic fibers, such as linear aromatic polyesters, with hydrophilic fibers, such as cellulose. Rubber compounders blended virgin natural rubber with reclaimed rubber for many years in order to reduce costs and, since 1940, have blended styrene-butadiene elastomer (SBR) with natural rubber in order to extend the supply of the latter.

When petroleum was inexpensive, there was not much incentive to blend general-purpose polymers. However, various polyolefins were blended to improve specific properties, and elastomers were blended with polystyrene to increase the impact resistance of this brittle polymer. These blends or alloys may be homogeneous with a single glass transition temperature (T_g) or heterogeneous with dispersed microphases having different T_g's.

In general, a negative change in free energy (ΔG) is essential for miscibility of polymer blends. The change in entropy (ΔS) is small enough to be negligible, and hence the change in enthalpy (ΔH) must be negative, zero, or very small, to ensure miscibility. The **Gibbs free energy equation** is shown below.

$$\Delta G = \Delta H - T\Delta S$$

Table 10–15. Thermal, Physical, and Chemical Properties of Typical Liquid-Crystal Polymers (LCP)

Property	Unfilled LCP(a)	50% Talc-filled LCP
Heat deflection temperature at 1820 kPa, °C	350	325
Maximum resistance to continuous heat, °C	250	250
Compressive strength, kPa	42,000	42,000
Flexural strength, kPa	125,000	110,000
Impact strength, Izod: cm · N/cm of notch	135	70
Tensile strength, kPa	135,000	70,000
Elongation, %	4.0	3.0
Hardness, Rockwell	R60	R76
Specific gravity	1.35	1.84
Dielectric constant	3	3.5
Water absorption, %	0	0
Resistance to chemicals at 25 °C:(b)		
Nonoxidizing acids (20% H_2SO_4)	S	S
Oxidizing acids (10% HNO_3)	S	S
Aqueous salt solutions (NaCl)	S	S
Aqueous alkalies (NaOH)	S	S
Polar solvents (C_2H_5OH)	S	S
Nonpolar solvents (C_6H_6)	S	S
Water	S	S

(a) Conversion tables appear in Appendix. (b) S, satisfactory; Q, questionable; U, unsatisfactory.

The requirement for a negative ΔH value is met when there is an attraction, such as hydrogen bonding or donor-acceptor attraction, between the two components of the blend. This requirement is met by polypropylene oxide-polystyrene blends as discussed in Section 10.8, by PVC-acrylonitrile-elastomer (NBR) blends, and by blends of polyvinylidene fluoride (PVDF), acrylates (such as polyalkyl acrylates and polyalkyl methacrylates), and LDPE with other polyolefins.

Some of the other commercial polymer blends consist of mixtures of incompatible polymers, such as high-impact polystyrene (HIPS) and ABS. The compatibilities of these component polymers may be improved by the production of graft copolymers.

The U.S. market for polymer blends and alloys was valued at over $450 million in 1983 and will approach $1 billion in 1988. The development of polymer blends or alloys is a cost-effective means to fill gaps in the performance of existing polymers, to improve processibility, heat resistance, toughness, and flame retardancy, and to increase sales without major capital expenditures or capacity expansion.

The leading common polymer blends — that is, polypropylene/ethylene-propylene terpolymer (16%), ABS/PVC (15%), modified nylon (10%), and polyphenylene oxide (PPO)/polystyrene (PS, 43%) — now account for over 80 percent of the total volume, but because of increasing demands for polymer blends, these "big four blends" will account for less than 70 percent in 1988.

Acrylonitrile-butadiene-styrene polymers (ABS) compete with polyvinyl chloride (PVC) for the extruded pipe market, but blends of these two resins are also widely used. PVC improves the flame retardancy, chemical resistance, and rigidity, and ABS provides improved processibility and impact resistance to the blend. Borg-Warner (Cycloloy) and Monsanto are the leading American suppliers of these PVC-ABS blends. General Electric, the world's largest supplier of engineering resins, supplies a blend of PVC and acrylic/styrene/acrylonitrile (ASA) called Geloy.

ABS/polycarbonate blends are supplied by Mobay (Bayblend) and Borg-Warner, and ABS/polysulfone blends (Mindel) are supplied by Union Carbide. Mindel has a heat transition temperature of 150 °C. Nylon/polyolefin blends are produced by duPont (Zytel), Allied (Capron), and Celanese. These blends have higher impact strengths, lower water absorption, and lower moduli than nylon.

In addition to the ABS/polycarbonate (PC) blends, PC has been blended with polyesters, styrene–maleic anhydride copolymers (Arloy), and polybutylene terephthalate (PBT, Xenoy). According to duPont, its polyacetal (POM/elastomer blend) is a true alloy seven times tougher than POM. The toughest polyethylene terephthalate (PET) has also been enhanced by blending with elastomers (Rynite).

Blends of PBT/PET are also available under the trade names Gafite, Celanex, and Veloy. Styrene–maleic anhydride (SMA) terpolymers, which are characterized by good resistance to heat, have been blended with PPO.

10.14 Future Directions for High-Performance Polymers

The future of high-performance polymers will be influenced by a demand for these unique materials and by new and improved products seeking a market. The growth of the engineering plastic segment of the polymer industry will depend on the extent to which they replace metal and ceramics in existing applications, on new fabrication techniques, and on the introduction of products with improved properties at a price that is in line with competitive materials.

The primary area for potential growth is transportation. However, large-scale replacement of metals in ground transportation will require lightweight, stiff engineering plastics with good dimensional stability, adequate strength, and good resistance to flame. Other areas for future growth of engineering plastics are as replacements for metal in appliances, power lines, and other industrial applications, and for applications in communications, electronics, and electrical industries.

The excellent physical and chemical properties of today's engineering polymers and those yet to be marketed ensure continued sound growth. This growth will be enhanced by the increased use of blends and composites. Most important, the sustained growth will depend on making the information on engineering polymers available to those who, until now, have not recognized the true potential of these unique materials of construction.

Physical and Chemical Testing of Polymers

11.1 Testing Organizations

The selection of general-purpose polymers has too often been the result of trial and error, misuse of case history data, and/or questionable guesswork. However, since high-performance polymers must be functional, it is essential that they be tested using meaningful, application-oriented procedures. Both the designer and the user should have an understanding of the testing procedure used in the selection of a high-performance polymer for a specific use. They should know both the advantages and the disadvantages of the testing procedure used and should not hesitate to develop additional empirical tests.

Fortunately, there are many standards and testing organizations whose sole purpose is to ensure satisfactory performance of materials. The largest of these is the International Standards Organization (ISO), which consists of members from 89 countries and many cooperative technical committees.

In addition, there are the American National Standards Institute (ANSI) and the American Society for Testing and Materials (ASTM), which publishes its tests on an annual basis in its Part 34, including committee reports from D20. Other important reports on tests and standards are published by the National Electrical Manufacturers Association (NEMA), Deutsches Institut für Normenausschuss (DIN), and the British Standards Institute (BSI).

11.2 Evaluation of Test Data

Unlike physical data available for metals, data for polymers depend on the life span of the test, the rate of loading, the temperature, the preparation of the test specimen, and other factors. Some, but not all, of these factors have been taken into account in obtaining the data tested in tables in previous chapters of this book. Published data may vary for the same polymer fabricated on different equipment, produced by different firms and for different formulations of the same polymer. Hence, the values cited in the tables were labeled "properties of typical polymers."

Many tests used by the polymer industry are adaptations of those developed previously for metals and ceramics. None are so precise that they can be used with complete reliability. In most instances, the physical, thermal, and chemical data are supplied by the producers, who are expected to promote their products in the marketplace. Hence, in the absence of other reliable information, positive data should be considered as upper limits of average test data and an allowance of at least ±5% should be made by the user or designer.

11.3 Heat Deflection Test (ANSI/ASTM D648-72/78)

This standard, now called Deflection Temperature of Plastics under Flexural Load (DTUL), is a result of "round-robin" testing by all interested members of ASTM Committee D20. This standard was accepted several decades ago. As indicated by the numbers after D648, it was revised and reapproved in 1972 and reapproved in 1978, respectively.

The test measures the temperature at which an arbitrary deformation occurs when molded or sheet plastics are subjected to an arbitrary set of testing conditions. The standard molded test span measures 127 mm long, 13 mm thick, and 3 to 13 mm wide. The specimen is placed in an oil bath under a load of 460 kPa or 1820 kPa in the apparatus shown in Fig. 11–1, and the temperature is raised at a rate of 2 °C/min.; the temperature is recorded when the specimen deflects by 0.25 mm.

Since crystalline polymers, such as nylon-66, have a low heat deflection temperature value when measured under a load of 1820 kPa, this test is often run at 460 kPa.

The results of this test must be used with caution. The established deflection is extremely small, and in some instances may be, at least in part, a measure of warpage or stress relief. The maximum resistance to continuous heat is an arbitrary value for useful temperatures, which are always below the DTUL value.

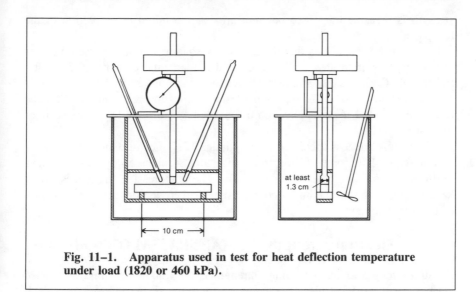

Fig. 11–1. Apparatus used in test for heat deflection temperature under load (1820 or 460 kPa).

11.4 Coéfficient of Linear Expansion Test (ANSI/ASTM D696-79)

Since it is not possible to exclude factors such as changes in moisture, plasticizer, or solvent content, and release of stresses with phase changes, ASTM 696 provides only an approximation of the true thermal expansion. Of course, the values for thermal expansion of polymers are high, relative to that of other materials of construction, but these values may be reduced by the incorporation of fillers.

In this test the specimen, measuring between 50 and 125 mm in length, is placed at the bottom of an outer dilatometer tube and below the inner dilatometer tube. The outer tube is immersed in a bath and the temperature is measured. The increase in length (ΔL) of the specimen measured by the dilatometer is divided by the initial length (L_0), multiplied by the increase in temperature in order to obtain the coefficient of linear expansion (α). The formula for calculating this value is shown below.

$$\alpha = \Delta L / L_0 T$$

11.5 Compressive Strength Test (ANSI/ASTM D695-77)

Compressive strength, also called compression strength, is the maximum stress that a rigid material will withstand under longitudinal compression. This strength is measured as force per unit area of the initial cross section of the test piece.

In the case of plastics that do not fail by shattering fracture, the compressive strength is an arbitrary value and not a fundamental property of the material tested. Compressive strength values are meaningless if the specimen is compressed into a flat disk. Additional tests, such as impact, creep, and fatigue tests, are essential if the application differs widely from the load-time scale of the test.

The standard test specimen is a cylinder 12.7 mm in diameter and 25.4 mm in height. The force of the compressive tool is increased by the downward thrust of the tool at a rate of 1.3 mm/min. The compressive strength is calculated by dividing the maximum compressive load by the original cross section of the test specimen.

11.6 Flexural Strength Test (ANSI/ASTM D790-71/78)

Flexural strength or cross-breaking strength is the maximum stress developed when a bar-shaped test piece, acting as a simple beam, is subjected to a bending force perpendicular to the bar. An acceptable test specimen is one that is at least 3.2 mm deep and 12.7 mm wide and long enough to overhang the supports, but the overhang should be less than 6.4 mm on each end.

The load should be applied at a specified cross-head rate, and the test should be terminated when the specimen bends or is deflected by 0.05 mm/min. The flexural strength (S) is calculated from the following expression:

$$S = PL/bd^2$$

in which P is the load at a given point on the deflection curve, L is the support span, b is the width of the bar, and d is the depth of the beam. A sketch of the test is presented in Fig. 11–2.

One may use the following expression:

$$r = 6Dd/L$$

in which D is the deflection to obtain the maximum strain (r) of the specimen under test.

One may obtain data for flexural modulus, which is a measure of stiffness, by plotting stress (S) vs. strain (r) during the test and measuring the slope of the curve obtained. The effect of the load (D) on the test bar in the ASTM Test 790 is shown by the sketch in Fig. 11–2.

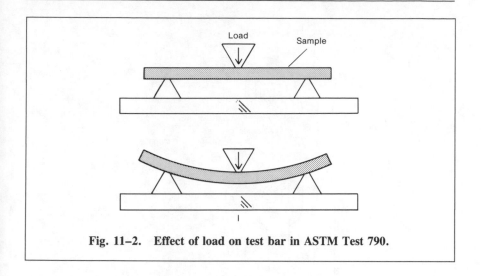

Fig. 11–2.   Effect of load on test bar in ASTM Test 790.

11.7 Impact Test (ANSI/ASTM D256-78)

Impact strength may be defined as toughness, or the ability of a rigid material to withstand a sharp blow, such as from a hammer. The information obtained from the most common test (ASTM D256) on a notched specimen, as shown in Fig. 11–3, is actually a measure of notch sensitivity of the specimen.

In the Izod test, a pendulum-type hammer, capable of delivering a blow of 2.7 to 21.7 J, strikes a notched specimen, measuring 63.5 mm × 1.27 m × 1.27 m with a 0.025-mm notch, which is held as a cantilever beam.

The distance that the pendulum travels after breaking the specimen is inversely related to the energy required to break the test piece, and the impact strength is calculated for a 25.4-mm test specimen.

11.8 Tensile Strength Test (ANSI/ASTM D638-77a)

Tensile strength or tenacity is the stress at the breaking point of a dumbbell-shaped tensile test specimen. The elongation or extension at the breaking point is the tensile strain. In Fig. 11–4, the test specimen is 3.2 mm thick and has a cross section of 12.7 mm. The jaws holding the specimen are moved apart at a predetermined rate and the maximum load and elongation at break are recorded. The tensile strength is the load at break divided by the original cross-sectional area. The elongation is the extension at break divided by the original gage length multiplied by 100. The modulus is the stress divided by the strain.

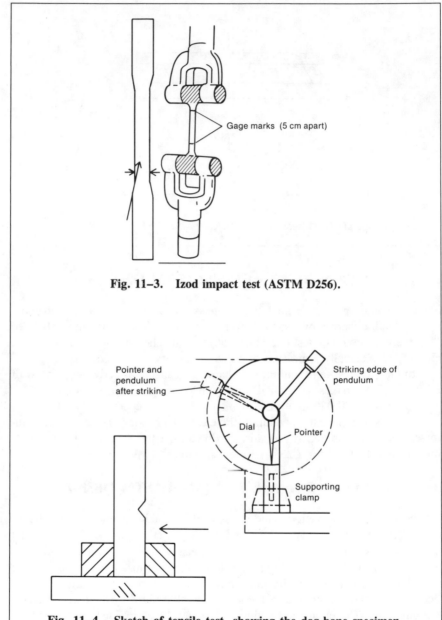

Fig. 11–3.   Izod impact test (ASTM D256).

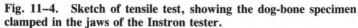

Fig. 11–4.   Sketch of tensile test, showing the dog-bone specimen clamped in the jaws of the Instron tester.

11.9 Hardness Test

Hardness is the resistance of a material to local deformation. The hardness test utilizes an indenter that may be a sharp pointed cone in the Shore D durometer test or a ball in the Rockwell test.

The Shore durometer is a spring-loaded indenter, with a scale that shows the extent of indentation with 100 being the hardest rating on the scale.

The Rockwell tester measures the indentation of a loaded ball, usually on the R scale. The diameter of the ball used in the R scale is 12.7 mm.

11.10 Density (Specific Gravity) Test

Density is usually expressed as grams per cubic centimeter (g/cm^3). Specific gravity is equal to the mass of a specific volume of the polymer compared to the mass of the volume of water. Since the g/cm^3 units for the plastic material and water cancel out, specific gravity has no dimensions.

11.11 Test for Resistance to Chemicals (ANSI/ASTM D543-67/78)

The resistance of polymers to chemical reagents has been measured by ASTM D543, which includes 50 different reagents. In the past, changes in weight and appearance of the immersed test sample have been reported. However, this test has been updated to include changes in physical properties as a result of immersion in test solutions.

Most high-performance polymers are not adversely affected by exposure to nonoxidizing acids and alkalies. Some are adversely affected by exposure to oxidizing acids, such as concentrated nitric acid, and all amorphous linear polymers will be attacked by solvents with solubility parameters similar to those of the polymer. Relatively complete tables have been published that show resistance of polymers to specific corrosives.

Terms and Symbols, Trade Names, and Bibliography

Terms and Symbols

A: Symbol for a repeating unit in the polymer chain

ABA copolymers: Block copolymers with three domains

Ablative polymer: A polymer that slows the ablative process

Abson: ABS

Accelerator: Catalyst for cross-linking

Acetal linkage: Oxygen linkage

Acetal polymers: $+CH_2O+$

Acetate rayon: Cellulose diacetate

Acrylamide: $H_2C=CHCONH_2$

Acrylic: Polymers based on acrylic acid esters; acrylonitrile or methacrylic acid

Acrylic fiber: Fiber based on polymers of acrylonitrile

Acrylon: Acrylic fiber

Acrylonitrile: $H_2C=CHCN$

Adipic acid: $HOOC(CH_2)_4COOH$

Addition polymerization: Chain polymerization resulting from addition of repeating units

Aldehyde: $RHC=O$

Alkanes: $H (CH_2)_n H$

Alkyd: Reactive product of phthalic anhydride, a glycol, and usually an unsaturated oil

Alkyl: $H(CH_2)_n$

Allophanate: Reactive product of an isocyanate and the hydrogen atoms in a urethane

Alloprene: Chlorinated rubber

Alternative copolymer: A copolymer in which each repeating unit is joined to another repeating unit in the polymer chain $(-A-B-A-B-)$

Amide group: $-CONH_2-$

Amilon: Nylon-6

Amine: RNH_2

Aminimide: Blocked isocyanate

Amino acid: $RCHNH_2COOH$

Amorphous nylon: Noncrystalline, transparent nylon

Anaerobic adhesives: Compounds that polymerize in the absence of oxygen

Anion: Negatively charged atom

Anode: Positive electrode

ANSI: American National Standards Institute

Antioxidant: Stabilizer against oxidation

Aramids: Aromatic polymers

Ardel: Polyarylate resin

Arsenic pentafluoride: AsF_5

Aspect ratio: Length/diameter

Atactic polymer: A polymer in which the pendant groups have a random arrangement on both sides of the polymer chain

Atactic polypropylene: Polypropylene with random arrangement of pendant groups

A-Tell: Polyester ether resin

ATH: Alumina trihydrate

Atom: The smallest distinct part of an element

Average molecular weight: The value obtained by dividing the sum of the molecular weights by the number of molecules present

B: Symbol for a repeating unit on the copolymer chain

Backbone: The principal chain in a polymer

Bakelite: Phenol formaldehyde resin

Balata: *Trans*-polyisoprene

Bayblend: ABS/PC

Bayer, O.: Inventor of polyurethane

Bifunctional reactants: Reactants with two functional groups

Bis-phenol-A: $(HOC_6H_4)_2C(CH_3)_2$

Bitumens: Asphalt-like polymers

Black Orlon: Pyrolyzed PAN fiber

Blendex: ABS

Block copolymer: A copolymer consisting of long sequences of each repeating unit, such as A_nB_n

BMC: Bulk molding compound

Bond, covalent: Chemical bond in which the electrons are shared

Bond, single: Bond made up of two shared electrons

Bond angle: The angle at which one atom is joined to another atom; this is 109.5° for C—C bonds

Bond length: The average distance between two atoms; this is 0.154 nm for C—C bonds

Boron: B

Branched polymer: Polymer in which chain extension takes place at more than one position on the polymer chain

BSI: British Standards Institute

Buna S: SBR

Butanol: $H(CH_2)_4 OH$

Butaprene-N: NBR

Butyllithium: $C_4H_9^-$, Li^+

Butyl rubber: Copolymer of isobutylene and isoprene

C: Carbon atom

c: Capacitance

Cab-o-sil: Finely divided silica

Calendar: Liquid-crystal polymer

Capron: Nylon-66

Carbanion: Negatively charged organic compound (ion)

Carbon monoxide: CO

Carbonium ion: Positively charged organic compound (ion)

Carboxylic acid: RCOOH

Catalyst: A substance that affects the rate of reaction but is not present in the product of the reaction

Catenated atoms: Atoms that are linked together, usually by covalent bonds

Cation: A positively charged atom

CBA: Chemical blowing agent

CED: Cohesive energy density

Celcon: Polyacetal

Celluloid: Plasticized cellulose nitrate

Cellulose xanthate: Reaction product of sodium cellulose and carbon disulfide

Ceramics: Materials based on baked clay

Chain-reaction polymerization: Addition polymerization

Chain transfer agent: A molecule from which an atom, such as hydrogen, may be readily abstracted by a free radical

Char: Coke formation

Chardonnet: Co-inventor of rayon

Chelate: Five- or six-membered ring formation based on intramolecular attraction of H, O, or N atoms

Chemical blowing agent: An agent that readily decomposes to produce a gas

Chemigum: SBR

Chlorinated rubber: Parlon

Cl: Chlorine

Cleavage: Breakage of covalent bonds

CMC: Carboxylmethylcellulose

Coagulation: Precipitation of a polymer dispersed in a latex

Coating: A layer of polymer, ceramics, or metal applied to a substance to protect it from the environment. Because of the presence of the substrate, this coating may be thin and not necessarily a self-supporting film.

Coefficient of expansion: Increase in volume per degree of temperature

Colligative properties: Properties based on the number of molecules present

Collodion: Solution of cellulose nitrate

Combing: Lining up of fibers

Composites: Polymers plus additives, usually reinforcements

Concentration gradient: dc/dy

Condensation polymerization: The condensation of two difunctional reactants, usually with the elimination of a small molecule

Conductive polymers: Polymers that conduct electricity

Configurations: Related structures produced by the cleavage and reforming of covalent bonds

Conformations: Different shapes of polymers resulting from rotation about single covalent bonds in the polymer chain

Coordination catalysis: Ziegler type of catalysis

Copolymer: Macromolecule made up of more than one repeating unit

Coumarone: Coal-tar resin

Coupling: The joining of two macroradicals to produce a larger molecule

Coupling agents: Difunctional compounds with attractive groups for fillers and resins

CPE: Chlorinated polyethylene

CR: Neoprene

CR-31: An allylic resin

Creep: Movement of a specimen under stress over a long period of time

Critical chain length: Minimum chain length required for entanglement

Creslan: Acrylic fiber

Cross: Co-inventor of viscose rayon

Cross-linked polymer: Three-dimensional polymer

Crylor: Acrylic fiber

Crystalline: Extent of crystals in a substance

Crystalline polymer: Any polymer containing crystalline areas

Crystals: Solids with characteristics based on the ordered arrangement of atoms or molecules in a definite pattern

Cumar: Coal-tar resin

Cumulative: Additive, or referring to the summation of effects

Cyanacryl: Polyacrylic elastomer

Cyclized rubber: Isomerized natural rubber

Cycloloy: PVC-ABS blend

D: Density

Darvan: Copolymer of vinylidene nitrile and vinyl acetate

D_c**:** Diffusion coefficient

Degree of polymerization: Number of repeating units in a polymer

Denier: Weight in grams of 9000 m of yarn

Density: Mass per unit volume

Derakane: Vinyl ester resin

Dexion: Polyester fiber

Diatomaceous earth: Finely divided silica from skeletons of diatoms

Diadic polyamide: Polyamide produced by the condensation of a diamine and a dicarboxylic acid

Dielectric strength: Maximum voltage that a polymer can withstand

Diffusion: Permeation of a gas from a higher to a lower pressure

Difunctional reactants: Reactants with functional groups

Diisocyanate: $R(CNO)_2$

Dilatometer: Device for measuring changes in volume

DIN: Deutsches Institut für Normenausschuss

Dimer: Oligomer with DP of 2, that is M_2

Disproportionation: Termination by chain transfer between macroradicals to produce a saturated and an unsaturated polymer molecule

DMF: Dimethylformamide

Dolan: Acrylic fiber

Dopant: An additive, such as AsF_5, that increases conductivity of polymers

DOP: Dioctyl phthalate

DP: Degree of polymerization; that is, the number of repeating units in a polymer molecule

Drawing: Extending or stretching

Drier: Organic salt of a heavy metal

Drying oil: Unsaturated vegetable oil

DS: Degree of substitution in cellulose

DTA: Diethylenetetramine

DUCO: Cellulose nitrate lacquer

Duprene: Neoprene

Durel: Polyarylate

Durez: Phenol formaldehyde resin

DVB: Divinylbenzene

E: Energy

Ebonite: Hard rubber

ΣG: Summation of Small's molar attraction constant

EEA: Ethylene-ethyl acrylate copolymer

EGG equation: (Einstein, Gold, Guth) $n/no = 1 + 2.5C + 14.1C^2$

Ekanol: Polyarylate

Ekonal: Polyarylate

Elastomer: A rubber-like polymer with high extensibility and low intermolecular force

Electromagnetic interference: Interference related to accumulated electrostatic charge in a nonconductor

Electrostatic charge: The accumulation of electricity in a nonconductor

Electron: A negatively charged lightweight particle

Electron, valence: An electron in the outer shell of an atom

Ellis, Carlton: Co-inventor of unsaturated polyester

EMF: Electromagnetic force

EMI: Electromagnetic interference

Emulsion polymerization: Polymerization of monomers dispersed in an aqueous emulsion

Enant: Nylon-7

End group analysis: The determination of molecular weight by determining the number of end groups present in the molecule

Enthalpy: H = heat content

Entropy: S = a measure of disorder, equal to zero at 0 K

EP: Epoxy resin

EPDM: Cross-linked ethylene-propylene copolymer

Epikote: Epoxy resin

Epoxy resin: Reaction product of *bis*-phenol-A and epichlorohydrin

Epoxylite: Epoxy resin

EPS: Expanded polystyrene

Epsilon (ε): Dielectric constant

EPR: Epoxy resin

Equilibrium: The state of forward and reverse forces being in balance

Ethanol: C_2H_5OH

Ethylene: $HC_2 = CH_2$

EVA: Ethylene-vinylacetate copolymer

EVAL: Ethylene-vinyl alcohol copolymer

Fabrication: Conversion of polymer to finished article

Fibers: Thread-like, crystalline materials with extremely strong intermolecular attractions, which favor good fitting of adjacent polymer chains

Fiberfax: Aluminum silicate fiber

Fiber K: Polycaprolactone

Fibrillation: Production of fiber from film

Fickian: Obeying Fick's law

Fick's law: $F = -D\dfrac{dc}{dx}$

Filament: Continuous fiber

Filament winding: The winding of a resin-impregnated filament on a mandrel

First-order transition: T_m = melting point

Flexural strength: Bending strength

Flory-Huggins equation: An equation for the partial molar Gibbs free energy of dilution

Folded chain: A crystalline polymer in which the polymer chain is folded back and forth

Fluorel: Polyfluorocarbon

Fumaronitrile:

Formica: Phenolic laminate
Fortrel: Polyester fiber
FRP: Fiber-reinforced polymer
Free energy: $\Delta G = \Delta H - T\Delta S$
Free radical: An electron-deficient molecule or atom
Free rotation: The rotation of atoms, particularly carbon atoms, about a single bond. Since the energy requirement is only a few kcal, the rotation is said to be free.
Full contour length: The length of a fully extended polymer chain

GPA: 145,038 psi
GPC: Gel permeation chromatography
g/denier: 0.0088 N/tex
Gel coat: Unfilled outer coat
Gel permeation chromatography: Liquid-solid chromatography technique used to separate polymer molecules according to their size
Gibbs, W.: Developer of phase rule and free energy concept
Gore-Tex: Microporous PTFE fiber
Graphite fibers: Carbon fibers
Grilene: Polyester fiber
Guayule: Desert plant with high rubber content
Gutta-percha: *Trans*-polyisoprene

H⁻: Hydride ion
H⁺: Proton
HALS: Hindered amine light stabilizer
Hardness: The quality of a substance to resist penetration or scratching
HDPE: High-density polyethylene; that is, linear polyethylene
HDTUL: Heat deflection temperature, under load
Head-to-head: Describes a configuration in which the functional groups are on

adjacent carbon atoms on a polymer chain
Head-to-tail: Describes a configuration in which the functional groups are as far apart as possible on a polymer chain
Heat deflection temperature: Temperature at which a loaded bar deflects a specified distance
Heat distortion temperature: Same as heat deflection temperature
Henry's law: Law by which the weight of dissolved gas is proportional to the pressure
HET anhydride: Chlorinated cyclic anhydride
Hevea braziliensis: Natural rubber
Hexafluoropropylene: $F_2C = C(CF_3)F$
Hexa: Hexamethylenetetramine
High-performance polymers: Engineering polymers
Hildebrand: Developer of solubility concept
Hindered phenols:

HIPS: High-impact polystyrene
Homologous: Referring to a series of organic compounds which differ by the number of methylene groups (CH_2)
Homopolymer: A polymer; name used to distinguish it from copolymers
Hooke's law: $S = G\gamma$: stress = modulus × strain
Horner: Fabricator of polymers
Hyatt, J. W.: Inventor of Celluloid
Hycar: NBR
Hydrocarbon: A compound consisting of carbon and hydrogen atoms, such as octane
Hydrogen bonds: The attractions between a hydrogen atom and an oxygen or nitrogen atom
Hydrogen cyanide: HCN
Hydrogenation: Addition of hydrogen to an unsaturated compound

Hydrolysis: Decomposition by water

Hydrophilic: Having an affinity for water

Hydrophobic: Having an aversion to water

Hylene: Blocked isocyanates

Hypalon: Sulfochlorinated polyethylene

Hytrel: Polyester thermoplastic elastomers

IIR: Butyl rubber

Index of refraction:

$$\frac{\text{sine of angle of incidence}}{\text{sine of angle of refraction}}$$

Initiation: The first step in polymerization reactions

Initiator: A substance that initiates a chain reaction

Inorganic: Noncarbonaceous compound

Instron: Instrument used for measuring tensile strength

Intermolecular attraction: Attraction between atoms in different molecules

Ion-exchange resins: Cross-linked polymers that form salts with ions from aqueous solutions

Ionomer: Copolymer of vinyl monomer and acrylic acid

IR: *cis*-Polyisoprene

ISO: International Standards Organization

Isobutylene: $H_2C:C(CH_3)_2$

Isonate: Blocked isocyanate

Isoprene:

$$H_2C = \overset{\overset{\textstyle CH_3}{|}}{C} - \overset{\overset{\textstyle H}{|}}{C} = CH_2$$

Isotactic polypropylene: Polypropylene with pendant groups on one side of the chain

Isotherm: Constant temperature line

it: Isotactic

Izod test: Impact test

J/M: cm $\cdot$ N/cm $= 0.0187$ ft·lb/in.

K: Kelvin

k: A constant in equations, such as the Mark-Houwink equation

Kalrez: Polyfluorocarbon

Kaolin: Clay

Kauri-butanol number: An empirical measure of solubility

kcal: Kilocalorie

Ketone: $R_2C{=}O$

Kienle: Inventor of alkyds

kPa: 0.145 psi

K factor: A measure of heat transfer

Kodel: Polyester fiber

Kraton: ABA block copolymer

Krynar: Polyacrylic elastomer

Kynar: Polyvinylidene fluoride

Ladder polymer: A polymer with two polymer chains

Lamellar: Plate-like in shape

Latex: Aqueous dispersion of a polymer

LDPE: Low-density polyethylene

Lexan: PC

Light scattering technique: The determination of molecular weight by measuring the light turbidity of a liquid system containing polymers

Linear: Describing a continuous chain

Lirelle: Polyester fiber

LDPE: Low-density polyethylene

LLDPE: Linear low-density polyethylene

LOI: Limiting Oxygen Index

London dispersion forces: Weak intermolecular forces based on transient dipole-dipole interactions

LP-3: Liquid Thiokol

Lycra: Polyurethane "snap-back" fiber

m: Molecular weight of a repeating unit

M: Molecular weight

M$^+$: Cation

M$^-$: Carbanion

M_c: Critical chain length

$\overline{M}_n$: Number-average molecular weight

$\overline{M}_w$: Weight-average molecular weight

Macrocarbonium ion: Positively charged macromolecule

Macromolecule: Extremely large molecule or polymer

Macroradical: Polymeric radical

Mark-Houwink equation: $[n] = k\overline{M}^a$

Makrolon: PC

Melmac: MF

Melting point: Temperature at which crystalline and liquid phase are in equilibrium

Mercerization: Alkaline treatment of cellulose

Merlon: PC

Metal: A substance characterized by thermal and electrical conductivity, luster, durability, and malleability. Metals consist of atoms with less than one-half the full complement of electrons in their outermost shells. The clusters of cations in metals are surrounded by a sea of loosely held electrons.

Methyl rubber: Poly 2,3-dimethylbutadiene

Methylcellulose: Methylether of cellulose

Methylene groups: $-CH_2-$

MF: Melamine-formaldehyde resin

Micarta: Phenolic laminate

Microballoons: Hollow spheres

Modacrylic: Modified acrylic fiber containing 35 to 85% acrylonitrile

Moduli: Plural of modulus

Modulus: Stress divided by strain minus stiffness

Molding compound: Mixture of polymer and additives ready for molding

Mole: Weight of a molecule in grams

Monadic polyamide: Polyamide produced from an amino acid

Mondur: Blocked isocyanate

Monodisperse polymer: A polymer in which the molecular weights of all molecules are identical

Monofunctional reactants: Reactants with a single functional group

Monomer: Building block for polymers

MPa: 145 psi

N: The number of molecules present in a sample

NBR: Acrylonitrile elastomer

NEMA: National Electrical Manufacturers Association

Neoprene: Polychloroprene

Network: Cross-linked polymer

Newton's law: $S = \eta \dfrac{d\gamma}{dt}$

Nitrocellulose: Incorrect name for cellulose nitrate

Nonoxidizing acids: HCl, H_2SO_4, etc.

Noryl: PPO

NR: Natural rubber

N/tex: 11.33 g/denier

Novolac: Condensate of phenol and formaldehyde under acid conditions

Nucleation: Crystallinity formation

Nyder: Nylon-6

Nylon: A synthetic polyamide; repeating units contain amide groups $(CONH_2)$

Nylon-6: Polycaprolactam

Nylon-66: Polymer obtained by the condensation of hexamethylenediamine and adipic acid

Nyrim: Nylon

O: Oxygen

Oleoresinous paints: Coatings based on polymerizable vegetable oils

Opacity: Property of being impervious to light

Optical methods of analysis: Photometric analytical technique

Organic: Carbonaceous compound

Organic metals: Conductive polymers

Orlon: Acrylic fiber

Oxidizing acids: HNO_3, H_2CrO_4, etc.

P: Extent of reaction in Carothers equation

PA: Polyamide

PAA: Polyacrylic acid

PAN: Polyacrylonitrile

Panlite: PC

Parel: Polyether elastomer

Parlon: Chlorinated rubber

Parthenium argentatum: Guayule bush

Parton-6: Nylon-6

Parylene: Poly *p*-xylylene

PBA: Physical blowing agent

PBI: Polybenzimidazole

PBT: Polybutylene terephthalate

PC: Polycarbonate

PEA: Polyethyl acrylate

Pebax: Nylon block copolymer

PEEK: Polyether ether ketone

PEI: Polyether imide

Pendant group: A group attached to the main chain, such as the methyl group in PP

Pentane: $H(CH_2)_5H$

Perfluorinated polymers: Polymers that are completely fluorinated

Pergut: Chlorinated rubber

Perlon D: PUR

Perlon U: Polyurethane

Permittivity: Dielectric constant

Peroxy compounds: Compounds containing $O-O$ linkage

PES: Polyether sulfone

PET: Polyethylene terephthalate

Petlon: PET

Petra: PET

Phase change: Change from gas to liquid or liquid to solid; that is, transition

Phenol: C_6H_5OH

Phillips catalyst: Catalyst based on CrO_3, supported on SiO_2/Al_2O_3

Phosgene: $COCl_2$

Phosphazene: Polymer with $-N=P$ backbone

Phosphorus pentachloride: PCl_5

Photoconductive polymers: Polymers that conduct electricity in the presence of light

Photoinitiators: Catalysts for ultraviolet-radiated polymerization

Photoresist: A polymer that degrades under controlled exposure, as in silicon chips

Physical blowing agent: A gas, such as a fluorocarbon

PI: Polyimide

PIA: Polyimide amide

Piccolyte: Turpentine resin

Piccopale: Petroleum resin

Piezoelectro polymers: Polymers that generate an electric current when compressed

Plasticizers: Flexibilizing additives

Plastics: Molecules that can be molded to produce useful rigid solids. These molecules have moderately strong intermolecular forces.

Plastisol: A dispersion of polymers in a liquid plasticizer

Pliolite: Cyclized (isomerized) rubber

Plywood: Laminate of wood and resin

PMMA: Polymethyl methacrylate

Poisson's ratio: $\gamma = \gamma_e/\gamma_w$

Polyacetylene: $(HC=CH)_n$

Polyacrylamide: $\left[\begin{matrix} CH_2-CH \\ | \\ CONH_2 \end{matrix}\right]_n$

Polyacrylonitrile: Polymer with the repeating unit $-CH_2CHCN-$

Polyamic acid: Soluble precursor to insoluble cross-linked polymers

Polyarylate: Aromatic polyester

Polydichlorostyrene:

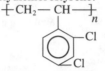

Poly 2,3-dimethylbutadiene: Methyl rubber

Polydisperse polymer: A polymer in which the molecular weights of the carbon molecules are different

Polydispersity index: $\overline{M}_w/\overline{M}_n$

Polyethyl acrylate: $\left[\begin{matrix} & H \\ & | \\ H_2C- & C \\ & | \\ & COOC_2H_5 \end{matrix}\right]_n$

Polyethylene: A polymer made up of repeating ethylene units $(-CH_2CH_2-)$ joined together in a long chain

Polyfluorocarbon: Fluorocarbon polymer, such as PTFE

Polyisobutyl ether: Maleic anhydride copolymer (Gantrez)

Polymer: An extremely large molecule consisting of a multitude of repeating units. The terms "giant molecule," "macromolecule," and "polymer" may be used interchangeably. The atoms in most polymers are joined by covalent bonds.

Polymer chain: The backbone of a polymer, usually consisting of a series of covalently bonded carbon atoms.

Polyphenylene sulfide:

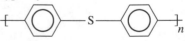

Polyphosphazene: Phosphonitrilic polymer

Polypropylene: A polymer with the repeating unit $-CH_2CH(CH_3)-$

Poly p-xylylene:

Polystyrene: A polymer with the repeating unit $+CH_2CH(C_6H_5)+$

Polysulfur nitride: $+SN\frac{}{n}$

Polyvinyl acetate: $+H_2C-CH\frac{}{n}$ | OOCCH$_3$

Polyvinyl alcohol: PVA, a polymer with the repeating unit $-CH_2CHOH-$

Polyvinyl chloride: PVC, a polymer with the repeating unit $-CH_2CHCl-$

Polyvinyl formal: Reaction product of PVA and formaldehyde

Polyvinylidene chloride: $+CH_2CCl_2\frac{}{n}$

POM: Polyacetal, polyoxymethylene

PP: Polypropylene

PPES: Polyphenyl ether sulfone

PPO: Polypropylene oxide

Prepolymer: Low-molecular-weight linear polymer

Pre-vex: PPO

Primary covalent bonds: Bonds between atoms

Processing: Compounding; that is, mixing of a polymer with additives

Propagation: Chain growth

Proteinaceous: Protein-containing

Proton: H$^+$

PS: Polystyrene

PTFE: Polytetrafluoroethylene

Pultrusion: The process in which a resin, impregnated in a bundle of filaments, is pulled through a heated die

PUR: Polyurethane

PVA: Polyvinyl alcohol

PVAC: Polyvinyl acetate

PVC: Polyvinyl chloride

PVCA: Polyvinyl carbazole

Pyrolysis: Degradation at high temperatures

Quiana: Polymer produced by the condensation of *bis-p*-amino-cyclohexyl-methane and dodecanedioic acid.

R: Ideal gas constant

R•: Free radical

Radical: Electron-deficient carbon atom

Random copolymers: Copolymers in which there is no definite order in the arrangement of the repeating units; that is, A—BAA—BA

Rayon: Regenerated cellulose from cellulose xanthate

RCOOH: Carboxylic acid

Reaction injection molding: A process in which polymerization occurs in the mold

Redon: Acrylic fiber

Refrasil: Silica fiber

Regenerated natural polymer: A polymer formed by the precipitation of fibers from a solution

Repeating unit: The monomeric unit in a polymer chain

Resol: Condensation of phenol and formaldehyde under alkaline conditions

Resol resin: Linear phenolic resin produced by alkaline condensate of phenol and formaldehyde

Reversible elongation: Recoverable elongation

Rilsan: Nylon-4

RIM: Reaction injection molding

RNH$_2$: Primary amine

Roylar: TPE

Rubber: A polymer of isoprene

Rucothane: TPE

Rust: Co-inventor of unsaturated polyester

S: Solubility coefficient

SAN: Styrene-acrylonitrile copolymer

Santocel: Finely divided silica

Saran: Vinylidene chloride polymer

Sb$_2$O$_5$: Antimony oxide

SBR: Styrene-butadiene rubber

Secondary valence bonds: Van der Waals forces

Segmental motion: Wriggling of polymer chain

Shore hardness: A measure of depth of needle penetration in polymers

Silanes: $H(SiH_2)_n H$

Silicone: Polysiloxane

Silk: Proteinaceous filament exuded by the silkworm

Siloxane: $+Si_2O+_n$

Sil-temp: Silica fiber

Simulated: Simplified formulas in which H atoms are not shown.

SMA: Styrene–maleic anhydride copolymer

Small's law: $n = \dfrac{\text{sine } i}{\text{sine } r}$

Small's relationship: Empirical equation for estimating solubility parameters

SMC: Sheet molding compound

Smith, Watson: Inventor of glyptals

Softening point: An empirical value related to T_g

Solprene: Styrene-butadiene block copolymer

Solubility: The extent to which a solute will dissolve in a solvent

Solubility parameter:
$$\delta = \left(\frac{D}{M}(\Delta H + RT)\right)^{1/2}$$

Spandex: Polyurethane "snap-back" fiber

Specialty polymers: Polymers that are useful for applications other than molding, extrusion, and thermoforming

Specific heat: Heat required to raise the temperature of one gram of material

Spectrophotometric methods of analysis: Photometric analysis and techniques

Spherulitic: Extent of aggregates of crystals present in a polymer

Spinneret: A series of small holes

Spinning: Passing through a small hole

SR: Synthetic rubber

Stanyl: Nylon-4,6

Starch: Poly α-D-glucose

Stereospecific: Polymers with specific arrangements of pendant groups in space, such as isotactic PP

Strain: Elongation resulting from applied stress

Stress: Force exerted on an object

Stymer: Water solution of SMA

Styrene: $CH_2CHC_6H_5$

Sucrose: A carbohydrate made of a D-glucose and a D-fructose unit, common name of sugar

Surlyn: Ionomer

Suspension polymerization: Polymerization of suspended beads of monomers in water

Swan, J. W.: Co-inventor of Rayon

Synergism: Cooperative effect of two additives

Syntactic foam: Polymer filled with hollow beads

TAC: Triallyl cyanurate

Tacticity: The arrangement of pendant groups in space

TDI: Tolylene diisocyanate

Tedlar: Polyvinyl fluoride

Teflon: PTFE

Teklan: Modacrylic fiber

Tensile strength: Resistance to pulling stresses

Tensile strength test: A test of resistance to pulling

Terephthalic acid: p-Phthalic acid

Termination: End of polymerization process

Tertiary amine: NR_3

Tertiary hydrogen atom:
$$-C-\underset{\underset{C}{|}}{\overset{\overset{H}{|}}{C}}-C-$$

Terylene: PET

Texan: Polyurethane

T_g: Glass transition temperature

Thermal conductivity: Conduction of heat

Thermal properties: Properties related to temperature

Thermoset: Cross-linkable or cross-linked

Thiokol: Polysulfide elastomer

T_m: Melting point, first-order transition

Ton, metric (t): 2204.6 lb

Torlon: Polyamide imide

Tornesit: Chlorinated NR

TPE: Thermoplastic elastomer

TPU: Thermopolyurethane elastomer

TPX: Polymethylpentene

Transdermal patch: Membrane that releases drugs at a controlled rate when adhered to the skin

Travis: Copolymer of vinylidene nitrile and vinyl acetate

Trevira: PET

UF: Urea-formaldehyde resin

Ultraviolet light stabilizers: Stabilizers against degradation in sunlight

Urylon: Polyurea

V: Volume

Valrin: Finely divided silica

Van der Waals forces: Attractive forces between molecules

Victroy: Polyphenyl sulfone

Veloy: PET/PBT blend

Verel: Modacrylic fiber

Vespel: PI

VI: Viscosity improver

Vicat softening point: Temperature at which a loaded needle penetrates a specified distance

Vinyl alcohol: A nonexistent monomer (H_2CCHOH)

Vinyl monomers: $\underset{\underset{H}{|}\ \underset{X}{|}}{C}=\underset{\underset{H}{|}\ \underset{H}{|}}{C}$, when X = Cl, COOH, C_6H_5, etc.

Vinylidene fluoride: $H_2C=CF_2$

Vinylite: Copolymer of vinyl chloride and vinyl acetate

Viscoelastic: Having the properties of an elastic solid and a viscous liquid

Viscosity index improver: Oil-soluble polymer

Viton: Polyfluorocarbon

Vitramid: Nylon-6

VLDPE: Very-low-density linear polyethylene

Vulcanization: Cross-linking, usually with sulfur

Vulcoprene A: PUR elastomer

Vydine: Nylon

Vyrene: PUR fiber

Wood flour: Attrition-ground wood

Wool: Proteinaceous fiber from sheep

Xydar: Liquid-crystal polymer

Yarn: Spun fibers

Zeolite: Aluminum silicate

Ziegler catalyst: Coordination catalyst based on $TiCl_3$ and ET_2AlCl

Trade Names

Aclar: Polyfluorocarbon

Aclon: Polyfluorocarbon

Acrilan: Acrylic fiber

Adiprene: Polyurethane elastomer

Alathon: Polyolefin

Alfane: Epoxy resin cement

Alkor: Furan resin cement

Amberlite: Ion-exchange resin

Ameripol: Polybutadiene

Amidel: Polyamide

Amproflex: PVC

Araldite: Epoxy resin

Ardel: Polyarylate

Arnite: PET

Astrel: Polysulfone

Bakelite: Phenolic resin

Barex: Acrylonitrile barrier resin

Bayblend: ABS/PC

Beetle: UF

Boltaron: PVC

Butacite: PVB

Butvar: PVB

Cadon: Styrene–maleic anhydride–acrylonitrile terpolymer

Capran: Nylon-6

Castimer: PUR

Celanar: Polyester

Celanex: PBT

Celcon: POM

Cellosize: Hydroxyethylcellulose
Chemigum: SBR
Cordura: Polyester fiber
Creslan: Acrylic fiber
Crofam: PMMA
Cumar: Coal-tar resin
Cyanoprene: Polyester polyol
Cycolac: ABS
Cymel: MF

Dacron: PET
Dapon: Polymer of allyl phthalate
Delrin: POM
Derakane: Vinyl ester resin
Desmodur: PUR
Dion: Unsaturated polyester
Dowex: Ion-exchange resin
Durel: Polyarylate
Durethane: Polyolefin
Durez: Phenolic resin
Dylan: Polyolefin
Dylark: Styrene polymer
Dylene: Polystyrene

Ekcel: Polyarylate
Ekonal: Polyarylate
Elvacite: PVAC
Epcar: Olefin-diene copolymer
Epolene: Polyethylene
Epotuf: Epoxy
Estane: PU
Ethocel: Ethylcellulose

Fiberglas: Fibrous glass
Fluorel: Fluoro polymers
Formica: PF laminate

Gafite: PET
Genal: PF
Geon: PVC
Grilon: Nylon

Halar: Polyfluorocarbon
Haveg: PF
Herculon: Polyolefin
Hetron: Unsaturated polyester
Hifax: Polyolefin
Hostalen: Polyolefin
Hypalon: Sulfochlorinated PE
Hytrel: Aromatic polyester

Isoplast: PU

Kalrez: Polyfluoro polymer
Kamax: Acrylic
Kapton: Polyimide
Kel F: Polyfluorocarbon
Kevlar: Aramid
Kodar: Cellulose acetate
Kodel: Polyester
Kodocel: Cellulose acetate
Korez: PF cement
Kralastic: ABS
Kraton: Styrene-butadiene block
K resin: Styrene copolymer
Kydene: PVC
Kydex: PVC
Kynar: PVDC
Kynol: PS

Lexan: PC
Lucite: Acrylic
Lustran: ABS
Lycra: Spandex PU

Marlex: Polyolefin
Marvinal: ABS
Melinex: Polyester
Merlon: PC
Minlon: Nylon
Mondur: PU
Moplen: PP
Mylar: Polyester

Natsyn: Polyisoprene
Neoprene: Polychloroprene
Norsidel: EPDM
Norsorex: Polynorboradiene
Noryl: PPO
Nyrin: Nylon-6

Orlon: Acrylic fiber
Oppanol: Polyisobutylene

Papi: Polyisocyanate
Parlon: Chlorinated rubber
Parylene: Polyphenylene
Paxon: PE
Pellothane: PU
Perspex: PMNA
Petlon: PET

Petra: PET
Petron: PET
Plaskon: UF
Plexiglas: PMMA
Pliofilm: Rubber hydrochloride
Pliovic: PVC
Pocan: PET
Pre-vex: Polyarylether
Profax: PP

Quacorr: Furan

Radel: Polyarylether
Rayslene: LPDM
Resumene: MF
Rilsan: Polyamide fiber
Rinthan: PU
Rovel: ABS
Rynite: PET
Ryton: PPS

Saflex: PVB
Santoprene: TPE
Saran: PVDC
Sayelle: PAN
Sclair: Polyolefin
Silastic: Silicone
Solprene: PS block
Somel: IPE
Styrofoam: PS foam
Styron: PS
Surlyn: Ionomer
Swedcast: PMMA

Tedlar: PVF
Teflon: PTFE
Tenlite: PTFE
Terylene: PET
Texin: Polyolefin
Torlon: Polyamide imide
Tufflex: PS
Turftane: PU
Tygon: Polyamide imide
Tynex: PU
Trevira: Polyester

Ucandel: Polysulfone
Udel: Polysulfone
Ultem: Polyamide imide
Ultrathene: Polyolefin
Urac: UF

Valox: PBT
Vespel: PI
Vibrathane: PU
Videne: PET
Vinylite: Polyvinyl chloride-co-vinyl acetate
Vistalon: PPDM
Vistanex: PIB
Viton: Polyfluorocarbon
Victrex: Polyether sulfone
Vulkollan: PU
Vydyne: Nylon

Xenoy: PC blend

Zytel: Nylon

Bibliography

Abraham, H. *Asphalts and Allied Substances*. Van Nostrand Reinhold, Princeton, NJ, 1960.

Ahmed, H. *Polypropylene Fibers—Science and Technology*. Elsevier Scientific Publishing, New York, 1982.

Allcock, H. R. *Phosphorus-Nitrogen Compounds*. Academic Press, NY, 1972.

Allcock, H. R. and F. W. Lampe. *Contemporary Polymer Chemistry*. Prentice-Hall, Englewood Cliffs, NJ, 1981.

Bailey, F. E., and F. V. Koleski. *Poly(ethylene oxide)*. Academic Press, NY, 1976.

Bauer, R. S. *Epoxy Resin Chemistry*. ACS Symposium Series 114, Washington, DC, 1979.

Becher, P. C., and M. N. Yudenfreund. *Emulsions, Latices and Dispersions.* Dekker, NY, 1977.

Bikales, N. M. *Mechanical Properties of Polymers.* Wiley-Interscience, NY, 1971.

Bikales, N. M., and L. Segal. *Cellulose and Cellulose Derivatives.* Wiley-Interscience, NY, 1971.

Billmeyer, F. W. *Textbook of Polymer Science.* Wiley-Interscience, NY, 1984.

Black, W. B., and J. Preston. *Aromatic Polyamide Fibers.* Dekker, NY, 1973.

Blackley, D. C. *High Polymer Latices.* Palmerton Publishing, NY, 1966.

Blumstein, A. *Liquid Crystalline Order in Polymers.* Academic Press, NY, 1978.

Blythe, A. R. *Electrical Properties of Polymers.* Cambridge University, UK, 1979.

Boenig, H. V. *Unsaturated Polyesters: Structure and Properties.* Elsevier, Amsterdam, 1964.

Boundy, R. H., R. F. Boyer, and S. M. Stroesser. *Styrene: Its Polymers, Copolymers and Derivatives.* Reinhold Publishing, NY, 1952.

Bovey, F. A., and F. H. Winslow. *An Introduction to Polymer Science.* Academic Press, NY, 1979.

Brighton, C. A., G. Pritchard, and G. A. Skinner. *Styrene Polymers: Technology and Engineering Aspects.* Applied Science Publications, London, 1979.

Briston, J. H., and L. L. Katon. *Plastic Films.* Wiley & Sons, NY, 1974.

Broutman, L. V., and R. H. Krock. *Composite Materials.* Academic Press, NY, 1974.

Broutman, L. V., and R. H. Krock. *Modern Composite Materials.* Addison-Wesley, Reading, MA, 1967.

Brydson, J. A. *Plastic Materials.* R. E. Krieger, Huntington, NY, 1975.

Buchter, H. H. *Industrial Sealing Technology.* Wiley & Sons, NY, 1979.

Burgess, R. N. *Manufacturing and Processing of PVC.* Macmillan, NY, 1982.

Cagle, C. V. *Adhesive Bonding Techniques & Applications.* McGraw-Hill, NY, 1968.

Carraher, C. E., J. E. Sheets, and C. U. Pittman. *Organo-Metallic Polymers.* Academic Press, NY, 1978.

Casey, J. P. *Pulp and Paper: Chemistry & Chemical Technology,* Wiley & Sons, NY, 1980.

Cassidy, P. E. *Thermally Stable Polymers, Synthesis and Properties.* Dekker, NY, 1980.

Ceresa, R. V. *Block and Graft Copolymers.* Butterworths, Washington, DC, 1962.

Chendraeskher, S. *Liquid Crystals.* Cambridge Univ. Press, London, 1977.

Christopher, W. F., and D. W. Fox. *Polycarbonates.* Reinhold Publishing, NY, 1962.

Cowie, J. M. G. *Polymers: Chemistry and Physics of Modern Materials.* Intex Educational Publishers, NY, 1973.

Crank, J., and G. S. Park. *Diffusion in Polymers*. Academic Press, London, 1968.

Danusis, A. *Sealants*. Reinhold Publishing, NY, 1967.

Davidson, R. L. *Water Soluble Gums and Resins*. McGraw-Hill, NY, 1969.

Deanin, R. D. *Polymer Structure and Applications*. Cahners Books, Boston, MA, 1972.

Doremus, R. H. *Glass Science*. Wiley & Sons, NY, 1973.

Dreyfuss, P. *Polytetrahydrofuran*. Gordon & Breach, NY, 1982.

DuBois, J. H., and F. W. John. *Plastics*. Van Nostrand, NY, 1960.

Eisenman, G. *Membranes*. Dekker, NY, 1975.

Elias, H. G. *Macromolecules: Structure and Properties*. Plenum, NY, 1977.

Erich, F. R. *Science & Technology of Rubber*. Academic Press, NY, 1978.

Ferry, J. D. *Viscoelastic Properties of Polymers*. Wiley & Sons, NY, 1970.

Finch, C. A. *Polyvinyl Alcohol*. Gordon & Breach, NY, 1970.

Flory, P. J. *Principles of Polymer Chemistry*. Cornell Univ. Press, Ithaca, NY, 1952.

Fox, D., and W. Christopher. *Polycarbonates*. Reinhold Publishing, NY, 1962.

Foy, G. F. "Engineering Plastics and Their Commercial Development." *Advances in Chemistry*, Vol. 96. Washington, DC, 1969.

Frados, J. *Plastics Engineering Handbook*. Van Nostrand Reinhold, NY, 1976.

Frazer, A. H. *High Temperature Resistant Polymers*. Wiley-Interscience, NY, 1968.

Frisch, K. *Electrical Properties of Polymers*. Technomic, Westport, CT, 1972.

Galan, A. V., and C. L. Martell. *Polypropylene Fibers*. Plenum, NY, 1965.

Gaylord, N. G. *Polyethers*. Wiley Interscience, NY, 1963.

Gaylord, N. G. *Reinforced Plastics*. Cahners Books, Boston, MA, 1974.

Ham, G. E. *Copolymerization*. Interscience, NY, 1964.

Ham, G. E. *Vinyl Polymerization*. Dekker, NY, 1969.

Harris, S. T. *The Technology of Powder Coatings*. Portcullis Press, London, UK, 1976.

Heifferich. *Ion Exchange*. McGraw-Hill, NY, 1962.

Hoiberg, A. *Bituminous Materials: Asphalts, Tars, and Pitches*. Wiley & Sons, NY, 1964.

Horn, M. B. *Acrylic Resins*. Reinhold Publishing, NY, 1961.

Hudlicky, M. *Chemistry of Organic Fluorine Compounds*. Wiley & Sons, NY, 1976.

Jenkins, A. D. *Polymer Science*. American Elsevier, NY, 1972.

Katz, H. S., and Milewski, J. V. *Handbook of Fiber and Reinforcements for Plastics*. Van Nostrand Reinhold, NY, 1978.

Kaufman, H. S., and J. U. Falcetta. *Introduction to Polymer Science and Technology*. Wiley & Sons, NY, 1977.

Kennedy, J. P., and E. Tornqvist. *Polymer Chemistry of Synthetic Elastomers*. Wiley & Sons, NY, 1968.

Kohan, M. I. *Nylon Plastics*. Wiley & Sons, NY, 1973.

Kunin, R. *Ion Exchange Resins*. R. E. Krieger, Huntington, NY, 1972.

Lal, J. *Elastomers and Rubber Elasticity*. ACS Symposium Series 193, 1982.

Lee, H., D. Stoffey, and C. Neville. *New Linear Polymers*. McGraw-Hill, NY, 1967.

Lee, H., and K. Neville. *Handbook of Epoxy Resins*. McGraw-Hill, NY, 1967.

Lenz, R. W. *Organic Chemistry of High Polymers*. Wiley & Sons, NY, 1967.

Lenz, R. W., and F. Ciardelli. *Preparation of Stereo-regular Polymers*. Reidel, Dordrecht, Netherlands, 1980.

Leonard, E. C. *Vinyl Monomers*. Wiley & Sons, NY, 1969.

Lubin, G. *Handbook of Composites*. Van Nostrand Reinhold, 1982.

Ludewig, H. *Polyester Fibers, Chemistry, and Technology*. Wiley-Interscience, NY, 1971.

Luskin, L. S., and E. C. Leonard. *Vinyl and Diene Monomers*. Wiley & Sons, NY, 1971.

MacDermott, C. P. *Selecting Thermoplastics for Engineering Applications*. Dekker, NY, 1984.

McGregor, R. R. *Silicones*. McGraw-Hill, NY, 1941.

MacKnight, W. V., F. E. Karasz, J. R. Fried. *Polymer Blends*. Academic Press, 1978.

Mandelkern, L. L. *An Introduction to Macromolecules*. McGraw-Hill, NY, 1972.

Margolis, V. M. *Engineering Thermoplastics*. Dekker, NY, 1985.

Mark, H. F., S. M. Atlas, and E. Cernia. *Man-made Fibers, Science and Technology*. Wiley-Interscience, NY, 1968.

Megson, N. J. L. *Phenolic Resin Chemistry*. Academic Press, NY, 1958.

Milby, R. V. *Plastics Technology*. McGraw-Hill, NY, 1973.

Miles, D. C., and J. H. Briston. *Polymer Technology*. Chemical Publishing, NY, 1979.

Mohr, J. G. *Handbook of Technology and Engineering of Reinforced Plastics/Composites*. Van Nostrand Reinhold, NY, 1973.

Moncrieff, R. W. *Man-Made Fibers*. Newnes-Butterworths, London, 1975.

Moore, G. R., and D. E. Kline. *Properties and Processing of Polymers for Engineering*. Prentice-Hall, Englewood Cliffs, NJ, 1984.

Morgan, P. *Glass-Reinforced Plastics*. Interscience, NY, 1961.

Mort, J., and G. Pfister. *Electronic Properties of Polymers*. Wiley & Sons, NY, 1982.

Morton, M. *Rubber Technology*. Van Nostrand Reinhold, NY, 1973.

Myers, R. R., and J. S. Long. *Treatise on Coatings*. Dekker, NY, 1972.

Nass, L. I. *Encyclopedia of PVC*. Dekker, NY, 1976.

Natta, G., and F. Danusso. *Stereoregular Polymers and Stereospecific Polymerization*. Pergamon, NY, 1967.

Nielsen, L. E. *Mechanical Properties of Polymers and Composites*. Dekker, NY, 1974.

Noll, W. *Chemistry and Technology of Silicones*. Academic Press, NY, 1968.

Odian, S. *Principles of Polymerization*. Wiley & Sons, NY, 1981.

Ogorkiewiz, R. M. *Thermoplastic Properties and Design*. Wiley & Sons, NY, 1974.

Ott, E., H. M. Spurlin, and M. W. Griffin. *Cellulose and Cellulose Derivatives*. Interscience, 1954.

Park, W. R. R. *Plastic Film Technology*. Van Nostrand Reinhold, NY, 1969.

Patten, W. J. *Plastics Technology*. Reston Publishing, Reston, VA, 1976.

Platzer, N. A. J. "Addition and Condensation Polymerization Processes." *Advances in Chemistry Series*, Vol. 91. ACS, Washington, D.C., 1969.

Plueddemann, E. P. *Composite Materials*. Academic Press, NY, 1974.

Plueddemann, E. P. *Silane Coupling Agents*. Plenum, NY, 1974.

Raff, R. A. V., and K. W. Doak. *Crystalline Olefin Polymers*. Interscience, 1965.

Raff, W. J., J. R. Scott, and J. Pacitt. *Handbook of Common Polymers*. Butterworths, London, 1971.

Renfrew, A., and P. Morgan. *Polyethylene: The Technology and Uses of Ethylene Polymers*. Interscience, NY, 1968.

Renner, E., and G. V. Samsonov. *Protective Coatings on Metals*. Plenum, NY, 1976.

Richardson, P. N. *Introduction to Extrusion*. Society of Plastics Engineers, Brookfield Center, CT, 1971.

Rochow, E. G. *Introduction to the Chemistry of Silicones*. Wiley & Sons, NY, 1951.

Rodriquez, F. *Principles of Polymer Systems*. McGraw-Hill, NY, 1971.

Rosato, D. N., and C. S. Grove. *Filament Winding*. Wiley-Interscience, NY, 1964.

Rosen, S. L. *Fundamental Principles of Polymer Materials for Practicing Engineers*. Cahners Books, Boston, MA, 1971.

Rubin, I. I. *Poly(1-butene): Its Preparation and Properties*. Gordon & Breach, NY, 1968.

Rubin, I. I. *Injection Molding Theory and Practice*. Wiley & Sons, NY, 1972.

Saltman, W. M. *The Stereo Rubbers*. Wiley-Interscience, NY, 1977.

Saunders, J. H., and K. C. Frisch. *Polyurethanes — Chemistry and Technology*. Wiley-Interscience, NY, 1962.

Schildknecht, C. E., and I. S. Skeist. *Polymerization Processes, High Polymers*. Wiley & Sons, NY, 1977.

Schlenkerh, B. R. *Introduction to Materials Science*. Wiley & Sons, NY, 1975.

Schnell, H. *Chemistry and Physics of Polycarbonates*. Wiley-Interscience, NY, 1964.

Schwartz, S. S., and S. H. Goodman. *Plastics Materials and Processes*. Van Nostrand Reinhold, NY, 1982.

Sears, J. K., and J. R. Darby. *The Technology of Plasticizers*. Wiley-Interscience, NY, 1981.

Seniegen, S. T., and C. S. Falk. *The Vanderbilt Rubber Handbook*. R. T. Vanderbilt, Roanoke, CT, 1978.

Seymour, R. B. *Introduction to Polymer Chemistry.* McGraw-Hill, NY, 1971.
Skeist, I. *Handbook of Adhesives.* Van Nostrand Reinhold, NY, 1977.
Society of the Plastics Industry. *Plastics Industry Safety Handbook.* Cahners
    Books, Boston, MA, 1973.
Staudinger, H. *From Organic Chemistry to Macromolecules.* Wiley & Sons,
    NY, 1963.
Swanson, R. S. *Plastics Technology.* McKnight & McKnight, Bloomington,
    IL, 1965.
Tess, R. W., and G. W. Pochlein. *Applied Polymer Science.* ACS, Washing-
    ton, DC, 1985.
Thorne, J. J. *Plastics Processing Engineering.* Dekker, NY, 1973.
Titow, W. V., and B. J. Lankem. *Reinforced Thermoplastics.* Applied Science
    Publishers, London, 1975.
Towiner, S. B. *Diffusion and Membrane Technology.* Reinhold Publishing,
    NY, 1962.
Ulrich, H. *Introduction to Industrial Polymers.* Carl Hanser Verlag Muncher,
    Germany, 1982.
Vale, C. P., and W. G. K. Taylor. *Amino Resins.* Iliffe Books, London, 1964.
Vandenberg, E. J. *Polyethers.* ACS, Washington, DC, 1975.
Vinson, J. R., and T. W. Chou. *Composite Materials and Their Use in Struc-
    tures.* Halsted Press, NY, 1975.
Walker, B. M. *Handbook of Thermoplastic Elastomers.* Van Nostrand Rein-
    hold, NY, 1979.
Ward, F. M. *Mechanical Properties of Solid Polymers.* Wiley-Interscience,
    1971.
Ward, I. M. *Structure and Properties of Oriented Polymers.* Halsted Press,
    NY, 1975.
Wessling, R. A. *Polyvinylidene Chloride.* Gordon & Breach, NY, 1977.
Whitby, B. *Synthetic Rubber.* Wiley & Sons, NY, 1973.
Whittington, L. R. *Whittington's Dictionary of Plastics.* Technomic Publishing
    Co., Stamford, CT, 1968.
Williams, H. T. *Polymer Engineering.* Elsevier Scientific, NY, 1978.
Yokum, R. H., and E. B. Nyquist. *Functional Monomers.* Dekker, NY, 1973.

Appendix: U.S./SI Units: Definitions and Conversions

Standard Symbols

C	= Celsius	m	= meter
g	= gram	N	= newton
Hz	= hertz	Pa	= pascal
J	= Joule	s	= second
K	= kelvin	t	= metric ton
kg	= kilogram		

Multiplicative Prefixes

Numerical value	Prefix	Symbol
10	deka	da
10^3	kilo	k
10^6	mega	M
10^9	giga	G
10^{-2}	centi	c
10^{-3}	milli	m
10^{-6}	micro	μ
10^{-9}	nano	n

Metric Conversions

To convert Metric system	To U.S. system	Multiply by
Length:		
$cm/cm/°C \times 10^{-5}$	$in./in./°F \times 10^{-5}$	0.555
millimeter (mm)	inch	0.0394
centimeter (cm)	inch	0.394
centimeter (cm)	foot	0.0328
meter (m)	foot	3.2808
Area:		
square millimeter (mm^2)	square inch ($in.^2$)	0.0016
square centimeter (cm^2)	square inch ($in.^2$)	0.155
square meter (m^2)	square foot (ft^2)	10.7639
Volume:		
cubic meter (m^3)	cubic foot (ft^3)	35.3147
m^3/kg	in^3/lb	27,680
Mass:		
gram (g)	pound (lb)	0.0022
kilogram (kg)	pound (lb)	2.2046
metric ton (t)	pound (lb)	2204.6
Force:		
$cm \cdot N/cm$	ft/lb/in.	0.0187
J/m	lb/in.	0.0187
N	lb/f	0.225
N/tex	g/denier	11.33
Density:		
g/cm^3	lb/ft^3	62.43
Temperature:		
°C	°F	1.8 °C + 32
K	°F	1.8 K − 459.67
Pressure:		
kPa	psi	0.145
MPa	psi	145
GPa	psi	145,038

Conversion of Tabulated Data to U.S. Units

Metric or SI units	Multiplied by	U.S. units
Heat deflection temperature at 1820 kPa, °C	0.555	at 264 psi, °F
Maximum resistance to continuous heat, °C	1.8 + 32	°F
Coefficient of linear expansion, cm/cm/ °C × 10^{-5}	0.555	in./in./°F × 10^{-5}
Compressive strength, kPa	0.145	psi
Flexural strength, kPa	0.145	psi
Impact strength, Izod: cm · N/cm of notch	0.0187	ft·lb/in. of notch
J/M	0.0187	ft·lb/in. of notch
Tensile strength, kPa	0.145	psi
Elongation, %	1	elongation, %
Specific gravity	1	specific gravity
Tenacity, N/tex	11.33	g/denier
Pa · S	10	poise

Index

Chemists

Baekeland, L. H., 29, 34, 50
Banks, 56, 58
Bartz, 128
Bayer, O., 73, 83, 98
Beadle, 109
Berzelius, 84
Bevan, 65, 109
Bischoff, 134
Boch, 45
Boeree, 128
Brown, 128
Burt, 103
Butlerov, 127
Carothers, W. H., 15, 56, 65, 66, 68, 79, 128, 134
Chardonnet, 65
Collins, 79
Conix, 139
Conrad, 81
Cross, 65, 109
Dickson, 68
Dufraisse, 39
Duppa, 62
Eareckson, 139
Edison, 77
Edmonds, 142
Einhorn, 134
Ellis, C., 37, 56, 94

Fawcett, 56
Fisher, C. H., 82
Fittig, 62
Fox, 134, 140
Franklin, 62
Friedel, 84
Gabriel, 66
Genvresse, 142
Gibson, 56
Goldschmidt, 51
Goodyear, C., 39
Haas, 62
Hammerich, 128
von Hedentron, 134
Henkel, 51
Hill, 142
Hogan, 56, 58
Holzer, 51
Hyatt brothers, 41
Ikeda, 104
John, 51
Kienle, 37, 93
Kipping, 84
Lenz, 142
Leuchs, 67
Levine, 139
Liebig, 85
Maass, 66

Subject Index